WADSWORTH PHILOSOPHERS SERIES

ON
KUHN

Hanne Andersen
University of Copenhagen

WADSWORTH
THOMSON LEARNING

Australia • Canada • Mexico • Singapore • Spain
United Kingdom • United States

COPYRIGHT © 2001 Wadsworth, a division of Thomson Learning, Inc. Thomson Learning™ is a trademark used herein under license.

ALL RIGHTS RESERVED. No part of this work covered by the copyright hereon may be reproduced or used in any form or by any means—graphic, electronic, or mechanical, including photocopying, recording, taping, Web distribution, or information storage and retrieval systems—without the written permission of the publisher.

Printed in the United States of America
1 2 3 4 5 6 7 04 03 02 01 00

For permission to use material from this text, contact us:
Web: http://www.thomsonrights.com
Fax: 1-800-730-2215
Phone: 1-800-730-2214

For more information, contact:
Wadsworth/Thomson Learning, Inc.
10 Davis Drive
Belmont, CA 94002-3098
USA
http://www.wadsworth.com

ISBN: 0-534-58356-3

Contents

PREFACE i

1.	BIOGRAPHICAL SKETCH	1
2.	PHILOSOPHICAL CONTEXT 8	
2.1	Logical positivism 8	
2.1.1	The movement of logical positivism 9	
2.1.2	Some core ideas 9	
2.1.3	Kuhn's problems with positivism 11	
2.2	The continental tradition 12	
2.3	The emergence of 'HPS' 14	
3.	THE STRUCTURE OF SCIENTIFIC REVOLUTIONS 18	
3.1	Genesis 18	
3.2	Main concepts of *Structure* 19	
3.2.1	Normal science 20	
3.2.2	Paradigm 22	
3.2.3	Anomalies 25	
3.2.4	Crisis and extraordinary science 27	
3.2.5	Revolution 29	
3.2.6	Mature science versus pre-paradigmatic science 32	
3.2.7	Incommensurability 32	
4.	SCIENTIFIC CONCEPTS 36	
4.1	The role of science teaching 37	
4.1.1	The aim of science education 38	
4.1.2	The means of science education 39	
4.2	A family resemblance account of concepts 41	
4.2.1	Concept acquisition 42	
4.2.2	Empirical vindication by computer simulation 44	
4.2	Philosophical problems 46	
4.4	Conceptual revolutions 48	
4.4.1	Kuhn and Quine on (un)translatability 49	
4.4.2	Conceptual microprocesses 51	
4.4.3	The scientific lexicon and anomalies as overlap 52	

5.	PHILOSOPHICAL IMPLICATIONS 55
5.1	Realism 56
5.1.1	The Kuhn-Putnam debate 56
5.1.2	Kuhn's non-realist stance 60
5.2	Truth 62
5.3	Rationality 64

6.	LATER DEVELOPMENTS 68
6.1	Concepts 68
6.1.1	Normic and nomic concepts 69
6.1.2	The biological component of the lexicon 71
6.2	The evolution metaphor 72
6.3	Historical evidence or first principles 75

NOTES 77

BIBLIOGRAPHY OF T. S. KUHN 89

BIBLIOGRAPHY 97

Preface

The present book *On Kuhn* is concerned with the philosophy of Thomas Samuel Kuhn (1922-1996). Kuhn's major work, *The Structure of Scientific Revolutions* from 1962 is one of the 20th century's most well-known philosophical works and has today become part of the canon, not only for philosopher, but for academics as such. Notions drawn from this work – most notably the notion of 'paradigms' – have become part of everyday speech, and for many the conception of paradigm shifts seem an obvious mode of thought, amounting almost to a truism.

The full significance of Kuhn's philosophy and the radicalism it presents may be seen most clearly on the background of the historical context in which it was produced. The first two chapters of this book are therefore devoted to a brief biographical account of Kuhn's development and career (chapter 1) and to a description of the reigning views in philosophy of science during the first half of the 20th century and the emergence of a historical philosophy of science by the mid-20th century (chapter 2).

Many of Kuhn's readers are acquainted only with his *The Structure of Scientific Revolutions*. However, *Structure* represents only the starting point of a developing philosophical position. The account of Kuhn's philosophy presented in this volume therefore starts with a detailed treatment of the content of *The Structure of Scientific Revolutions* (chapter 3) and proceeds to explain how his position developed into a complex account of the nature of scientific concepts (chapter 4) and account for the philosophical implications of Kuhn's ideas (chapter 5). The book closes with an exposition of the latest developments in Kuhn's thinking from the late 1980s and early 1990s in which he started departing from his previous views on a number of key points (chapter 6).

Intended to be of use both at the introductory and at more advanced levels, the basic presentation of Kuhn's position as well as the most important objections against it are covered in the main text. For the interested reader further details and references can be found in the footnotes.

Preface

Kuhn was trained first as a physicist, but made his career in history of science and philosophy of science. His academic production covers a wide field: solid state physics, the history thermodynamics, the history of optics, the chemical revolution, the Copernican revolution, the early history of quantum physics, historiography of science and philosophy of science. Focusing on Kuhn as a philosopher, this book primarily focuses on his philosophical publications. Further, stressing the importance of the possibility for the reader to trace the arguments presented in this book by readings of Kuhn's work, apart from a brief description of his Lowell Lecutres Kuhn's unpublished manuscripts are not included in present account of Kuhn's position.[1] A full bibliography of Kuhn's publications is provided at the end of the book. To facilitate the access to his publications the bibliography includes information of reprints. Publications which have been reprinted are in the text referred to both by the original year of publication and the publication year of the reprint. Unless otherwise indicated pagenumbers will refer to the most recent edition of the work.

Thanks are due to Nancy Nersessian, Peter Baker, Stig Brorson, Henriette Jørgensen and Carsten Sestoft who patiently read and commented on various drafts of this book. I am also grateful to Charles Gillispie for historical information, to Paul Hoyningen-Huene, Howard Sankey and Skuli Sigurdsson who provided me with various sources, and to Karl Hufbauer who kindly gave me access to his historical work on the young Kuhn.

1
Biographical Sketch[1]

Thomas Samuel Kuhn was born on July 18, 1922 in Cincinnati, Ohio, as the first child of Samuel L. and Minette Stroock Kuhn. The family moved to New York City shortly after Thomas was born, and another child, Roger, was born a few years later. Samuel L. Kuhn had been trained as a hydraulic engineer in a joint program at Harvard and MIT, but worked for a bank investigating investment possibilities. Minette Stroock did professional editing. The family was a liberal, Jewish family, forming part of the local intelligentsia. For the first nine years of his schooling, Thomas was sent to schools which were part of the progressive education movement, and one in particular was considered extremely liberal and left oriented. Later, he recalled that these schools had emphasized subject matter less than they emphasized independence of mind, and that, consequently, he had had little drill, but had learned early to do a good deal of work by himself. For the tenth to twelfth grades he was sent first to a school in Pennsylvania, then to a preparatory school in Connecticut. These schools offered more formal training, but he missed the sort of interactions he had had at his former school, and he did not particularly like the new schools.

In high school he had been very good at mathematics, and in 1940 he enrolled at Harvard to study physics. In the fall of his sophomore year the Japanese attacked Pearl Harbor, and the United States joined the war. As a member of the editorial staff at the Harvard *Crimson*, Kuhn wrote editorials in support of Harvard's president James Conant's (1893-1978) efforts to militarize the nation's colleges. As a result of the war, the physics department where he was enrolled changed largely to training the students in electronics, and he decided to take his degree as quickly as possible, graduating after only three years by attending summer schools and reducing the number of courses outside the sciences. After graduation he went to work for the Radio Research Laboratory investigating radar counter measures. He first spent a year at the laboratory in Boston, but was then sent to England and later to France, and he returned to Boston in the early summer of 1945.

Biographical Sketch

With the war ending, Kuhn started graduate work in physics at Harvard. While an undergraduate Kuhn had "wanted to find out about philosophy" (Kuhn et al. 1997, p. 151) and had attended a course in the history of philosophy, giving a presentation on Kant and the notion of preconditions for knowledge; a notion which "just knocked [him] over" (Kuhn et al. 1997, p. 152).[2] Starting as a graduate student he received permission from the physics department to use half of his first year studying philosophy. But Kuhn was much too independent-minded or perhaps too impatient to start in a new field from scratch. He later recalled that he "took two coureses, and I realized that there was just a lot of philosophy which I hadn't been taught, and didn't understand, and was not finding it very palatable to pick up this way" (Kuhn et al. 1997, p. 159). Rather than starting taking undergraduate courses in philosophy he decided to finish his degree in physics. In 1946 he passed the General Examinations and started Ph.D. research in theoretical solid-state physics.

While still a graduate student, Kuhn was invited by Conant in 1947 to assist in the development of a general-education course on science for Harvard juniors and seniors. The course was based on the close examination of a few historical case studies from the development of science; an idea which had been advanced in the recent report from the Harvard Committee on the Objectives of a General Education in a Free Society (published 1945) as well as Conant's own just-published book *On Understanding Science: An Historical Approach* (1947).

Kuhn willingly accepted Conant's invitation, and was asked to do a case study on the history of mechanics; an assignment that became decisive for Kuhn's intellectual development. He started studying the discussions of motion in Aristotle's *Physica*, approaching the text from a modern viewpoint. From that perspective much of what Aristotle had said seemed to him simply wrong, or even absurd. But Kuhn was puzzled. In other areas Aristotle had been a very gifted thinker – how could his talents have failed him only when it came to mechanics? In Kuhn's later work, he repeatedly recalled this 'Eureka experience': how one summer day he suddenly perceived an *alternate* way of reading Aristotle – a way in which Aristotle's claim suddenly made sense.[3] This experience dramatically influenced Kuhn's view on the nature of science. While discovering history, he had discovered his first scientific revolution.

By the end of the first semester Kuhn knew what he wanted to do. He was interested in philosophy but had already decided that he did not want to go back to taking undergraduate courses. Now, an alternative seemed to be to go into history and make philosophy out of it. But he was about to finish a dissertation in physics, not in the history of science. He therefore needed a career strategy that would enable him to make a drastic change of research field. A fellowship at the Harvard Society of Fellows would give him the peace to work himself into the new field, and Kuhn bluntly asked Conant to sponsor him for the Society.

During these years Kuhn had started psychoanalysis. He was not very happy with his analyst whom he later recalled to have fallen asleep, snoring, during the sessions, but he continued the analysis for a couple of years. Despite his discontents with the analyst, the experience of the analysis taught him a certain technique – the technique of understanding other people, "to climb into other peoples' heads" as he occasionally called it (Kuhn et al. 1997, p. 165) – which he adopted for his historical work.[4]

In 1948 Kuhn received a three-year fellowship. He finished his dissertation during the fall and began working in the Society of Fellows. The fellowship relieved him of teaching obligations and other responsibilities and gave him a period of freedom which he could devote to his transition from physicist to historian. He was already acquainted with the works of the philosopher and historian of science Alexandre Koyré (1892-1964), and he now added the philosopher of science Emile Meyerson (1859-1933), the historian of chemistry Hélène Metzger (1889-1944), and the historian of science Anneliese Maier (1905-1971) to his favourites – all scholars whom he thought had "shown what it was like to think scientifically in a period when the canons of scientific thought were very different from those current today" (Kuhn 1970a, p. vi). However, he did not confine his reading to the history of science, on the contrary, he undertook an apparently non-systematic study of a range of fields. Encouraged by a colleague he studied some psychology of perception, especially the Gestalt psychologists. Another colleague made him study the linguist Benjamin L. Whorf (1897-1941). Accidentally a footnote in a paper by the sociologist Robert Merton made him aware of the work of the psychologist Jean Piaget (1896-1980) which he studied carefully. Yet another footnote made him aware of the medical doctor and microbiologist Ludwik Fleck's (1896-1961) *Genesis and Develoment of a Scientific Fact*, and in addition he studied some formal logic, and so forth.

In Kuhn's third year as a fellow, Conant decided to stop teaching the general-education course on science, and Kuhn took over together with a colleague. While Conant had been giving the course, lots of people attended it, wanting to see the president of the University. When

Biographical Sketch

Conant left, enrollment dropped dramatically. Kuhn started getting nervous in advance of his teaching, and spent much time preparing – a nervousness he later claimed never to have quite gotten over (Kuhn et al. 1997, p. 168).

In 1951 he was asked to give the Lowell Lectures – a series of public lectures which had previously been given by quite distinguished speakers. Kuhn had "a dreadful time preparing it and ... nearly cracked up" (Kuhn et al. 1997, p. 172). But he had not set himself an easy task. He was only 29, he had not yet published as a historian of science, nor as a philosopher – and yet he had set out to deliver a series of lectures with the aim of showing that the reigning view of science was altogether wrong. This was the birth of the basic ideas of his *Structure of Scientific Revolutions* – but delivering the lectures primarily made him realize that neither his ideas nor his mastery of the historical cases were yet ripe for publication.

After the fellowship Kuhn got an appointment first as instructor and then as assistant professor at Harvard. His assignment was primarily the general education course, but he gradually started giving history of science courses. He was also elected to the History of Science Society's Council, and publications only appeared slowly, despite a one-year fellowship from the Guggenheim Foundation.

As a result, he could foresee difficulties obtaining tenure at Harvard, and he started looking for alternatives. In 1956 he accepted an appointment as assistant professor in history of science at Berkeley; a position which was split between the philosophy department and the history department. His first book, *The Copernican Revolution: Planetary Astronomy in the Development of Western Thought* was published the following year. This was a subject he had lectured on previously and had planned to write a book on during his Guggenheim fellowship. Finally he had managed to get it done, and it paved his way for tenure at Berkeley.

Having obtained tenure, Kuhn accepted a long-standing invitation from the Center for Advanced Study in the Behavioural Sciences at Stanford, California. His intention was to write the draft of *Structure* which had been on his mind now for a decade, but again, he was too optimistic. He soon produced a chapter on revolutionary change, but a companion chapter on the normal interlude between revolutions gave him great trouble. However, staying in a community composed predominantly of social scientists, he was surprised by the number and extent of the overt disagreements between social scientists about the nature of legitimate problems and methods. Attempting to discover the source of the difference between the social scientists' overt disagreement and natural scientists' overall agreement on the fundamentals of their science, Kuhn recognized the importance of the universally recognized scientific achievements that natural scientists adopt as model problems and solutions. This concept of 'paradigms' proved to be the missing piece of his puzzle, and a first full draft of *Structure* was prepared between the summer of 1959 and the end of 1960.

Kuhn had seen *The Copernican Revolution* as a book that "repeatedly violates the institutionalized boundaries which separate the audience for 'science' from the audience for 'history' or 'philosophy' (Kuhn 1957a, p. viii); *Structure* marked his transition from "relatively straightforward historical problems back to the more philosophical concerns that had initially led [him] to history" (Kuhn 1970a, p. v).

But whereas Conant had provided him with the fellowship necessary for the smooth transition from physicist to historian, the road from history back to his main interest, philosophy, proved more rocky. Kuhn returned from the Center in 1959. He spent the following year finishing a manuscript which he saw as the return to his original philosophical interest. Having received an offer from Johns Hopkins in the fall of 1960, he negotiated a promotion to full professor with Berkeley as a condition for staying. He declined the offer from Johns Hopkins, received a promotion to full professor in Berkeley – and then learned that the philosophers had decided to exclude him from their department and recommend his promotion to be only in history. Kuhn was "very deeply hurt"; the hurt never altogether went away (Kuhn et al. 1997, p. 182).

Simultaneously, he had been asked by the American Physical Society to direct an archival project on History of Quantum Theory, and in 1962 the Kuhn family moved for a year to Copenhagen, Denmark, which served as the European base of the project. While in Copenhagen he received an offer from Princeton. Two years previously Princeton had announced the creation of a special program of graduate studies in the history and philosophy of science which should enable students to pursue work in both fields. When the program was announced the staff consisted of two philosophers of science, Carl G. Hempel (1905-1997) and Hilary Putnam (1926-), and two historians of science, Charles Gillispie (1918-) and John Murdoch (1927-), with Gillispie as head of the program. Shortly upon his return from Europe Kuhn visited the program, decided to join them, and left Berkeley for Princeton in 1964.

Meanwhile, in 1962, *The Structure of Scientific Revolutions* had been published. Much to Kuhn's surprise he realized that it was receiving more response from social scientists than from philosophers.[5] Its popularity was steadily increasing, and during the student rebellions of the late sixties it was highly popular – but to Kuhn's dismay largely for the wrong reasons. As he later recalled:

> Students used to come to me saying things like "thank you for telling us about paradigms, now that we know what they are we can get along without them". All seen as examples of oppression. That wasn't my point at all! I remember being invited to attend and talk to a seminar at Princeton organized by undergraduates during the times of troubles. And I kept saying "But I didn't say that. But I didn't sayt that! But I didn't say that!" And finally, a student of mine, or a student in the Program who had sort of helped get me into this, and had come along to listen, said to the students, "You have to realize that in terms of what you are thinking of, this is a profoundly conservative book." And it is... (Kuhn et al. 1997, p. 188).

His production after *Structure* was both within history of science and philosophy of science – but he always kept the two subjects seperate, sometimes to the puzzlement of his readers.[6] However, it was his firm conviction that history and philosophy of science ought to remain two separate and distinct disciplines. Thus, in a lecture held in 1968 while he was chairman of the Princeton Program in History and Philosophy of Science, he insisted that "there is no such field" as the history and philosophy of science (Kuhn 1977b, p. 4). Though he felt the need for a dialogue between the two fields, it had to be interdisciplinary rather than intradisciplinary (cf. Kuhn 1977b, p. 4)

In 1978 he published the monograph *Black-Body Theory and the Quantum Discontinuity, 1894-1912*. This was a purely historical work, and some critics expressed disappointment that Kuhn's philosophical views were nowhere explicitly to be found. However, Kuhn maintained that although many of his philosophical notions originated in his experience writing history, he did his best

> not to think in these terms when I do history, and I avoid the corresponding vocabulary when presenting my results. It is too easy to constrain historical evidence within a predetermined mold. If history (or ultimately philosophy) is to be learned from the texts that are my main sources, then I must minimize the role of prior conviction in my approach to them. Often I do

Biographical Sketch

not know for some time after my historical work is completed the respects in which it does and does not fit *Structure* (Kuhn 1984a, p. 245).

By the end of the 1970s Kuhn received an offer from MIT which he accepted in 1979 and stayed until his retirement in 1991. In 1982 he was appointed the first Laurance S. Rockefeller professor in philosophy at MIT, and from 1990 to 1991 he was the elected President of the Philosophy of Science Association.

During the 1980s and 90s most of his work was devoted to the philosophical problems left over from *Structure of Scientific Revolutions*: rationality, relativism, realism, and truth (cf. Kuhn 1991a, p. 3). This work led to the publication of a number of papers, many of them on such issues as reference, meaning change and translation failure, but a projected book remained unfinished.[7] Thomas Samuel Kuhn died on June 17, 1996 after having been ill for two years with cancer.

2
Philosophical Context

Kuhn's philosophical work can be seen as a continuous struggle to develop a historical philosophy of science, that is, a philosophy of science that is informed by the history of science in its claims on the nature of science and that approaches science as a continuously developing, historical product (cf. Kuhn 1992). This emphasis on history places Kuhn's philosophy of science in a complicated philosophical context. As he put it: "Anyone who believes that history may have deep philosophical import will have to learn to bridge the longstanding divide between the Continental and English-language philosophical traditions" (Kuhn 1977a, p. xv).

Kuhn's own work pertains to issues derived both from Anglo-Saxon and Continental philosophy. Within Anglo-Saxon philosophy Kuhn's work is closely related to the discussions of the 1950s and 1960s on the problems of logical positivism (chapter 2.1). Whereas logical positivists made a clear distinction between the logical analysis of science as a topic for philosophy and the empirical investigations of science as a topic for history, sociology and psychology, it was common within the Continental tradition, especially in France, to see philosophy of science and history of science as closely related (chapter 2.2). Philosophers' criticism of logical positivism combined with historians' inspiration from the French-Russian philosopher Alexandre Koyré led to the development of an Anglo-Saxon historical philosophy of science; a development that was facilitated by institutional developments caused partly by the reform of general science education, partly by the recognition of history, philosophy and sociology of science as fields of enquiry that could be of use to science policy (chapter 2.3).

2.1 Logical positivism

Logical positivism (or, logical empiricism) is the name of a philosophical movement covering several related but somewhat different positions. The movement originated in Austria and Germany,

Philosophical Context

but developed to include a substantial number of philosophers from both Europe and the United States (chapter 2.1.1). At the core of the movement lie a number of claims regarding the meaning of scientific statements and the relation between theories and observation (chapter 2.1.2) – assumptions which have often been depicted in a somewhat distorted, simpleminded way (chapter 2.1.3).

2.1.1 The movement of logical positivism

Logical positivism developed from the Vienna Circle formed by the philosopher Moritz Schlick (1882-1936) in the early 1920s.[1] In 1929 three of the core members of the Circle, Carnap, Neurath and Hahn, published *The Vienna Circle: Its Scientific Outlook* which gave a brief account of the Circle's philosophical position and the problems within the philosophy of science with which it was primarily concerned.

The Vienna Circle early formed an alliance with the Berlin Society for Empirical Philosophy.[2] With the ambitions of developing an international movement, they organized a series of congresses beginning in 1929 which brought together philosophers from Scandinavia, the Netherlands, Germany, Poland, Great Britain and the United States. By the early 1930s, logical positivism had become an influential philosophical movement internationally. Logical positivists were not only interested in philosophy, but were also involved in contemporary political and cultural currents. Their ideas were progressive and liberal. Some members, such as Neurath, were outspoken socialists. The advent of Nazism therefore forced the majority of the Circle's members into exile, and several members of the Circle found new homes in the United States. In 1938 Neurath, Carnap and the American pragmatist Charles Morris began editing the *International Encyclopedia of Unified Science*. The *Encyclopedia* was planned like an onion in which the heart would be the two introductory volumes entitled *Foundations of the Unity of Science*. These two volumes were planned to consist of 20 pamphlets – of which Kuhn's *The Structure of Scientific Revolutions* appeared as issue 2, volume 2 in 1962.

2.1.2 Some core ideas

Determining criteria for demarcating between science and

Philosophical Context

metaphysics was a central concern of logical positivists. According to them, philosophy should focus solely on logical analysis as an activity through which the meaning of statements was revealed or determined. In doing this, philosophy should serve as a method for clarifying meaningful concepts and propositions, laying the logical foundation for science and mathematics, and establishing the empirical basis of science. In viewing the role of philosophy as laying the foundation for science by clarifying its concepts and propositions, the aim of logical positivism was not totally unlike the project which Immanuel Kant (1724-1804) had developed more than a century before. In his transcendental philosophy Kant had developed a notion of synthetic a priori knowledge, claiming that we can obtain certain knowledge about, for example, time, space, and causality independently of our experience. Kant claimed that the pure part of science, covering Euclidean geometry, Galilean kinematics and the Newtonian laws of motion, consisted solely of such synthetic a priori knowledge and therefore possessed apodictic certainty.

However, the development of non-Euclidian geometry and the theory of relativity in the late 19th and early 20th centuries showed that Kant's view was untenable. The logical positivists therefore rejected the Kantian notion of the synthetic a priori. Hence, only two sources of knowledge could be possible: logical reasoning (analytic a priori) and experience (synthetic a posteriori). Since the two sources of knowledge were supposed to be exhaustive, any meaningful sentence had to express either something that was formally true or something that could in principle be completely verified by observational evidence, otherwise the sentence had to be considered literally meaningless. This claim became known as the verifiability principle. The verifiability principle made meaning and verification co-extensive.

The verifiability principle was soon attacked, most notably by Popper (1901-1994). In his 1935 *Logic of Scientific Discovery* Popper argued that since the laws of natural science take the form of general sentences, these laws cannot be derived from observational evidence which takes the form of particular sentences. Any attempt to do so encounters the problem of induction, formulated by David Hume (1711-1776): generalizations about phenomena cannot be derived from any number of individual observations. Thus, since the laws of natural science cannot be completely verified by observational evidence, according to the verifiability principle they should therefore be seen as meaningless. To Popper, logical positivism rejected not only metaphysics, but natural science as well. His solution was to introduce the criterion of falsifiability as providing the demarcation between science and metaphysical speculation. According to this criterion, scientific statements are, in principle, falsifiable, while those of metaphysics are not. This criterion was, in turn, criticized by the positivists, though we will not discuss the details here. The logical positivists' solution to this problem was to require that for a sentence to be meaningful its constituent non-logical terms had to have an empirical basis, which for them meant being reducible to observational terms. This development started an important liberalization of logical positivism leading to a conventionalist stance that Carnap, in particular, elaborated. The nature of conceptual frameworks and the possibility of alternative frameworks were central issues of Carnap's research. Carnap treated scientific conceptual frameworks as languages comprising observational and theoretical vocabularies, rules of sentence formation, and rules for connecting the two vocabularies. As his position developed, he emphasized that the form of this system never was completely settled by experience but was always partially determined by conventions. Moreover, if the form of two systems providing alternative conceptualizations of a domain differed in essential respects, translation from the one language to the other might not always preserve the factual content of empirical statements. For example, many statements of modern physics cannot be translated completely into statements of classical physics. By the same token, a change in language could be a radical alteration, or in other words, a scientific revolution.

2.1.3 Kuhn's problems with positivism

Logical positivism included a wide range of philosophers and developed over several decades. As a movement it did not form a coherent whole, instead, some members would hold conflicting views, just as some views would change over time. However, the canonical view, not the more nuanced position, was influential for the development of Kuhn's philosophy of science and for the development of the historical philosophy of science generally.

In canonical form, logical positivism has often been depicted as characterized by a number of fixed principles: a clear distinction between a neutral observation language and the theoretical language, a strict principle of verifiability according to which only statements that have direct implications for sensory experience are meaningful, and a radical empiricism according to which sensory experiences are systematized by scientific theories that, in turn, develop to encompass more and more observable facts.

Late in his life Kuhn admitted that "it was against this sort of everyday image of logical positivism – I didn't even think of it as logical empiricism for a while – it was against that I was reacting when I saw my first examples of history" (Kuhn et al. 1997, p. 186).[3] This has caused some confusion in the treatments of the relations between Kuhn and logical positivism. Since Kuhn had indicated that he saw his early works as a 'decisive transformation in the image of science by which we are now possessed' (cf. Kuhn 1970a, p. 1), he was seen for a long time as one of the rebels attempting to refute logical positivism.

However, the logical positivists who had invited Kuhn to publish in their *Encyclopedia* did not consider Kuhn to be their adversary. Upon receiving the first draft of his manuscript, Carnap acknowledged to Kuhn:

> I believe that the planned monograph will be a valuable contribution to the Encyclopedia. I am myself very much interested in the problems which you intend to deal with, even though my knowledge of the history of science is rather fragmentary. Among many other items I liked your emphasis on the new conceptual frameworks which are proposed in revolutions in science, and, on their basis, the posing of new questions, not only answers to old problems (Carnap to Kuhn on April 12, 1960, quoted from Reisch 1991, p. 266).

Similarly, late in his life Kuhn stated that "I have confessed to a good deal of embarrassment about the fact that I didn't know [Carnap's view]. On the other hand, it is also the case that if I'd known about it, if I'd been into the literature at that level, I probably would never have written [*The Structure of Scientific Revolutions*]. And the view that emerges in *Structure* is not the same as the Carnap view, but it's interesting that coming from what were partially different ... Carnap staying within the tradition had been driven to this – I had rebelled already and come to it from another direction, and in any case we were still different" (Kuhn et al. 1997, p. 186).[4]

2.2 The continental tradition

In his writings as well as in interviews Kuhn referred to a variety of European historians and philosophers. Kant's notion of the

Philosophical Context

categories as preconditions for knowledge had "knocked him over" as a student (cf. Kuhn et al. 1997, p. 152), and Kantian ideas came to play an important role in the later development of his position which he sometimes characterized as Kantian, only with changing categories (cf. Kuhn 1979b, p. 418f, 1991a, p. 12. See also chapter 5.1). During the first half of the 20th century, several European philosophers had directed their work towards rejecting Kant's absolute categories in favor of categories that may change with history. For example, the German philosopher Ernst Cassirer (1874-1945) from the 'Marburg school' of neo-Kantianism rejected Kant's idea of the absolute a priori to develop instead a notion of the relative a priori.[5]

In France, the historian and philosopher of science Gaston Bachelard (1884-1962), student of the leading French neo-Kantian Léon Brunschvicg (1869-1944), noted that what Kant had taken to be absolute preconditions for knowledge had turned out wrong in the light of modern physics. On Bachelard's view, what had seemed to be absolute preconditions for knowledge was instead merely contingent conditions. Bachelard concluded that an account of scientific reason could only be derived from reflections upon its historical development. Based on the analysis of the historical development of science, Bachelard advanced a model of scientific change according to which occasionally the conceptions of nature are replaced by radically new conceptions – what Bachelard called epistemological breaks. Bachelard's view was later developed and modified by the historian and philosopher of science George Canguilhem (1904-1995) and by the philosopher and social historian Michel Foucault (1926-1984). Foucault described his project as an analysis of the history of thought. He compared his project to Kant's critique of reason – but with the difference that his interest was with a *historical* a priori, that is, with what seems in a given period to be necessary conditions governing reason and how these constraints have a contingent historical origin.

However, Kuhn's interest in European scholarship was not with neo-Kantianism or its offshoots.[6] Most of the Continental scholars to whom Kuhn referred – the most important being Alexandre Koyré (1892-1964) and Emile Meyerson (1859-1933), but also Hélène Metzger (1889-1944), Anneliese Maier (1905-1971), and occasionally Gaston Bachelard and Leon Brunschvicg – he admired mainly for their historical work, seeing them as pioneers in exploring a historical period on its own terms (see also chapter 2.3). Hence, he noted that "the early models of the sort of history that has so influenced me and my *historical* colleagues is the product of a post-Kantian European tradition which I and my *philosophical* colleagues continue to find opaque" (Kuhn 1977a, p. xv). Of Meyerson and Brunschvicg he declared that although he often urged his students to read them he recommended "these authors for what they saw in historical material not for their philosophies, which I join most of my contemporaries in rejecting" (Kuhn 1977b, p. 11).

The only exception may have been Bachelard. Late in his life Kuhn reported meeting Bachelard in Paris in 1950 when he had read very little of his work. As Kuhn recalled, due to language difficulties the two did not talk for long, and "it is perhaps a pity, because although I think I have read a bit more of the relevant material since, and have real reservations about it, nevertheless he was a figure who was seeing at least some of the thing. He was trying to put it in too much of a constrain ... He had categories, and methodological categories, and moved the thing up an escalator too systematically for me. But there were things to be discovered there that I did not discover, or not discover in that way" (Kuhn et al. 1997, p. 169).

2.3 The emergence of 'HPS'

The historical philosophy of science (often abbreviated HPS) as it emerged in the Anglo-American tradition during the late 1950s and early 1960s had its origin both in a new historiography of science inspired by the French tradition, and in the criticism of logical positivism.

History of science began to be established as an independent academic discipline during the late 19th and early 20th century. New professorships for the history of science were established, the first international conferences were held, and new societies and journals dedicated to the history of science were formed. In the United States, courses in the history of science had been offered at a few universities since the end of the 19th century. In 1924 the History of Science Society was formed, and Harvard University was the first American university to offer degrees in the history of science from the late 1930s.

During the 1940s and 1950s, history of science came to play an important role in the 'general science' syllabus in the United States. In contrast to the British or the German university education, the U.S. had developed a university system in which the first degree provides a general education, and students were therefore required to take courses from both the humanities and the sciences. However, during the first half of the 20th century there was an increasing discontent with the outcome of the science courses for non-science majors. To remedy this,

Philosophical Context

the Harvard Committee on the Objectives of a General Education in Free Society recommended in their 1945 report that general science courses should be characterized by broad, integrative elements such as the relation of science with its own past and human history in general, or comparisons of scientific with other modes of thought. At Harvard University, James B. Conant who was a strong advocate of the historical case-study approach to general science education, and himself trained as a chemist, started implementing the approach together with two assistants: the young chemist Leonard K. Nash and a graduate student in physics, Thomas S. Kuhn.

By the mid-1950s, the National Science Foundation began seeing history, philosophy and sociology of science as fields of enquiry that could help providing a better understanding of such issues as scientists' motivation, creativity, communication and decision making, the social organization of scientific institutions, the role of scientific societies, etc. In 1957 this led to the creation of the History and Philosophy of Science Program that should provide resources to train new scholars. From the mid-1940s to the mid-1960s, several history of science programs were established, a few in the form of combined history and philosophy of science programs.

During this period the relation between history and philosophy of science changed dramatically. According to the logical positivists, the history of science was of no particular interest to the philosophy of science. Their clear distinction between the logical reasoning of philosophy and the empirical knowledge obtained in the sciences could be found again in their distinction between the logical analysis of how scientific claims are justified and the empirical analysis of how they are discovered. This distinction became known as the distinction between the context of justification and the context of discovery.[7]

Whereas history of science was of little interest to philosophy, philosophy was clearly of use to history. George Sarton (1884-1956), the father of history of science in the United States, advocated a historiography according to which past achievements should admittedly be evaluated in relation to their predecessors. However, the standards according to which it was evaluated whether these past achievements constituted a step forward should be the modern standards of progress and rationality. On this perspective, philosophy of science could inform the historical analysis by providing the standards by which to judge earlier achievements. Similar views were expressed by other scholars such as, for example, the British historian of science and medicine Charles Singer (1876-1960).

However, another key figures for many of the young scholars

graduating during the late 1940s and early 1950s was the Russian-French philosopher Alexandre Koyré whose *Galileo Studies* provided a close study of the conceptual changes in astronomy and mechanics from Copernicus to Newton. From him a new generation of scholars learned to study past scientific ideas from the viewpoint that gives these ideas the maximum internal coherence and to display the historical integrity of these ideas in their own time.

By the late 1950s and early 1960s, a new movement arose that merged history with philosophy of science. The new historiography of science that attempted describing past science in its historical integrity rather than through the lens of the present showed an image of the scientific enterprise that seemed different from the image entailed by the standard philosophical accounts. Simultaneously, logical positivism had become subject of severe criticism. Among the critics were Norwood Russell Hanson (1924-1967) who argued against the distinction between the context of discovery and the context of justification (e.g. Hanson 1958), and Paul Karl Feyerabend (1924-1994) who attacked the canonical positivist view of reductionism (e.g. Feyerabend 1962/1981). The work of philosophers such as Hanson and Feyerabend suggested that philosophy of science should attend to the history and the actual practice of science if it was to be prevented from turning into abstract logic with little relation to the actual scientific enterprise. Driven both by historians investigating scientific reason through its historical development, and by philosophers questioning the ahistorical assumptions of the accepted philosophical viewpoints, the new ideal developed that philosophy of science should be concerned with the historical structure of science rather than with an ahistorical, logical structure that had proved to be a mere chimera.

This transition to what is often called the 'historical philosophy of science' is a transition for which Kuhn felt that he got "far more credit, and also more blame, than I have coming to me. I was, if you will, present at the creation, and it wasn't very crowded. But others were present too: Paul Feyerabend and Russ Hanson, in particular, as well as Mary Hesse, Michael Polanyi, Stephen Toulmin, and a few more besides. Whatever a *Zeitgeist* is, we provided a striking illustration of its role in intellectual affairs" (Kuhn 1991a, p. 3).

In attending to the actual scientific practices, the emerging historical philosophy of science was faced with the question whether the accounts it provided were descriptive or prescriptive. Kuhn recognized the question, but argued that his theses "should be read in both ways at once" (Kuhn 1970c, p. 237). According to Kuhn, his account of the development of science was a prescriptive theory, and

Philosophical Context

the reasons for taking it seriously were that scientists do in fact behave as the theory says they should (cf. Kuhn 1970a, p. 208). To avoid being accused of circularity, Kuhn emphasized that "the consequences of the viewpoint being discussed are not exhausted by the observations upon which it rested at the start" (Kuhn 1970a, p. 208). Like for any other theory, the success of a theory of the development of science should be dependent on its ability to explain new data that had not been involved in its initial formulation.[8]

3
The Structure of Scientific Revolutions

Kuhn's most famous work is the monograph *The Structure of Scientific Revolutions*. Originating in the Lowell Lectures held in 1951, the first full draft was completed between the summer of 1959 and the end of 1960. In 1962 it was published both as a separate book and as part of the *International Encyclopedia of Unified Science* (chapter 3.1). In 1970 a second edition of *Structure* was published which contains a postscript in which Kuhn addressed the main points of criticism that had been raised against his views.[1] *The Structure of Scientific Revolution* has been translated into more than thirty languages, and the book has sold nearly one million copies world-wide.

The main objective of the book is to question the view of scientific history as a cumulative development in which scientists gradually add new pieces to the ever-growing aggregate of scientific knowledge. Contrary to this view, Kuhn claimed that science develops through successive periods of tradition-preserving normal science and tradition-shattering revolutions (chapter 3.2).

3.1 Genesis

In March 1951, Kuhn held a series of eight lectures at the Lowell Institute. These lectures were his first attempt to spell out the position that science – understood as the activity in which scientists engage, rather than as a body of developed laws and techniques – is guided by certain orientations or preconditions which mark out what scientific problems and their solutions may look like. Hence, in doing science, scientists do not simply proceed from objective experimental facts to unique, immutable laws, instead, theories and facts are provided together by the scientific 'orientation'.

During the lectures Kuhn expressed his conviction that science progresses by a series of repeated attempts to apply differing points of view to the natural world. He claimed this to be not simply a contingent, historical fact; it was the nature of science to develop in such a pattern.[2] Almost half of the lectures were devoted to various historical case studies; the rest were an attempt to vindicate his view by arguments drawn from logic, linguistics, and psychology.

Around 1953 Kuhn was commissioned to write the history of science volume for the *International Encyclopedia of Unified Science*, and he planned the bulk of this volume to grow out of the Lowell Lectures. However, during the following decade his scholarly production was mainly historical, and his philosophical ideas were put to one side. During this period he published on 17th century chemistry (Kuhn 1952a), 18th and 19th century thermodynamics (Kuhn 1955a, 1955b, 1958a, 1960, 1961b), Newton's optics (Kuhn 1958b), and Aristotelian physics and the Copernican revolution (Kuhn 1957a). He also wrote a substantial number of reviews of books on the history of science and general intellectual history for the History of Science Society's journal *Isis*. By the end of the 1950s he was finally able to devote his full attention to his philosophical ideas again. While finishing the manuscript for the book, he published two papers on the convergent activity of normal research and the rigorous and conservative science education necessary for establishing convergent thought (Kuhn 1959a/1977a, 1961a/1977a); themes that would become essential for his arguments in *Structure* (see also chapter 4.1).

3.2 Main concepts of *Structure*

"History, if viewed as a repository for more than anecdote or chronology, could produce a decisive transformation in the image of science by which we are now possessed" (p. 1). With this line Kuhn opened *The Structure of Scientific Revolutions*, indicating that his aim was twofold. First, the history of science should be used to arrive at a different image of the development of science than had been seen previously, but second, in order to obtain this goal a new historiography had to be deployed.

In the introduction to *Structure* Kuhn described how the previous historiography had prescribed a history of science that merely chronicled the successive increments towards the present and the obstacles that had impeded this accumulation. The questions which this kind of historiography prescribes are questions such as when a given scientific fact was discovered and by whom, and which errors had prevented a more rapid development towards the present stand.

However, according to Kuhn these questions are surprisingly hard to answer – so hard, in fact, that he suspected that they are simply the wrong questions to ask (cf. p. 2). Hence, rather than asking questions derived from the present, Kuhn advocated a historiography that focused on the historical integrity of science at a particular time of its development. Thus, instead of describing the cumulative development *towards* a specific point in history, the present, history of science should see science as developing onwards *from* a given point in history.

Kuhn claimed that the image of science implied by this new diachronic historiography would differ fundamentally from the image implied by the previous, anachronic historiography.[3] Rather than a purely cumulative development, Kuhn described the development of science as successive periods of cumulative normal science separated by non-cumulative revolutions. This account of the development of science described *normal science* (chapter 3.2.1) as dependent on some set of received beliefs, a *paradigm*, which marks out what the acceptable research problems are and what acceptable solutions to these problem must look like (chapter 3.2.2). Yet, some of the scientific problems defined by a paradigm may turn out to be unsolvable within the framework of the paradigm, that is, turn into *anomalies* (chapter 3.2.3). If such anomalies cannot be resolved, they may cause a *crisis* in the scientific community (chapter 3.2.4), and this crisis may eventually lead to a *revolution* (chapter 3.2.5). The paradigms separated by such a revolutionary divide will in some areas be so different that the relation between the new paradigm and the old cannot be seen simply as one of extension or refinement. Kuhn termed this relation *incommensurability* (chapter 3.2.7).

This phase development of science, paradigm → crisis → revolution → new paradigm etc., requires that a paradigm has been established, that is, that the scientific community has reached a consensus on various fundamentals. This has happened at different times in history for the various scientific disciplines, hence, a distinction can be made between the *mature science* in which shared paradigms have been developed, and their *pre-paradigmatic* ancestors which are characterized rather by their variety of competing schools of thought (chapter 3.2.6).

3.2.1 Normal science

After his introductory chapter on the role for history in understanding science, Kuhn opened the next chapter with a definition of normal science as "research based firmly upon one or more past scientific achievements, achievements that some particular scientific community acknowledges for a time as supplying the foundation for its success" (p. 10).

According to Kuhn, research of this kind is not aimed at calling forth new sorts of phenomena or at inventing new theories (cf. p. 24), but solely at increasing the success of the accepted theory. Hence, normal science research usually falls within three classes: determination of significant facts, matching of facts with theory, or articulation of theory (cf. p. 34). Kuhn called these kinds of research 'puzzle-solving' (p. 35ff.), or 'mopping up operations' (p. 24).[4] The analogy between puzzle-solving and normal science is based on two similarities. First, both puzzles and research problems have an assured solution. The scientist who sets out, for example, to determine a physical constant is assured that this constant exists and can somehow be measured. Second, that certain rules must be obeyed in achieving the solution characterizes both puzzles and research problems.[5] In normal science these rules are, for example, the scientific laws provided by the accepted theory (p. 40), commitments to preferred types of instrumentation (p. 40), or metaphysical beliefs about what kinds of objects nature does and does not contain (p. 41).

These rules provide a set of tools accepted by and shared among the members of the scientific community, and in using these tools the scientists expect the solutions of new problems to be similar to the solutions of problems previously examined within the discipline. An alleged solution to a scientific puzzle that implicitly violates one of the rules, for example a natural law, is usually not considered an acceptable solution.

Kuhn argued that the confident and continuous use of the accepted tools enables science to move *faster* and penetrate *deeper* than if the tools were continuously changed. The restricted vision of normal science "forces scientists to investigate some part of nature in a detail and depth that would otherwise be unimaginable" (p. 24). Kuhn argued that the puzzle-solving character of normal science implies that scientific achievements are measured by solved problems. Thus, as long as the provided tools are used faithfully, it is well-defined which problems have already been solved and which problems should be solved to contribute further to the progress of science. By contrast, if the scientists change their tools by adopting a different theory, embodied in different rules, they may open up a whole range of new problems that need to be solved, or even reopen problems that had previously been considered solved. Thus, retooling will usually appear as a retrogression. As Kuhn remarked: "as in manufacture so in science – retooling is an extravagance to be reserved for the occasion that demands it" (p. 76).

3.2.2 Paradigm

To describe the consensus within a scientific community regarding both past exemplary achievements and future expectations of how to model research on these achievements, Kuhn introduced the concept 'paradigm'. According to the preface, Kuhn developed the concept in trying to grasp the difference he had observed between the social sciences and the natural sciences. Whereas the natural science communities do not have much controversy over fundamentals, the social sciences seemed to show a striking number of overt disagreement about the nature of legitimate scientific problems and methods (cf. p. viii). According to Kuhn, the source of this difference is that recognized scientific achievements in the natural sciences play the role of *exemplary* problems and problem solutions. Whereas agreement upon such exemplary problems and problem solutions had been achieved in all or most research disciplines within the natural sciences, Kuhn saw it as "an open question what parts of social science have yet acquired such paradigms at all" (p. 15. See also chapter 3.2.7).

In the paper 'The Essential Tension: Tradition and Innovation in Scientific Research?' (Kuhn 1959a/1977a) published while he completed the manuscript of *Structure*, Kuhn had introduced the term 'paradigm' to denote standard examples, intending its usage to be similar to how the term was used within language teaching (see chapter 4.1.2). Likewise, in *Structure* he first used the term to denote past scientific achievements in the form of the famous classic texts of science that had been both "sufficiently unprecedented to attract an enduring group of adherents away from competing modes of scientific activity" and "sufficiently open-ended to leave all sorts of problems for the redefined group of practitioners to resolve" (p. 10).[6]

However, later in *Structure* his use of the notion became ambiguous. Kuhn widened the notion considerably, claiming that contrary to the paradigms used in language teaching, in science "a paradigm is rarely an object for replication. Instead, like an accepted judicial decision in the common law, it is an object for further articulation and specification under new or more stringent conditions" (p. 23). Thus, Kuhn used the term paradigm both to denote concrete exemplary problems and problem solutions which have previously been achieved, and to denote a more loose or open-ended structure that marks out further problems to solve and the means by which to solve them. In some places this latter use of the term was carried further to denote the entire global set of commitments shared by the members of the scientific community.[7]

Kuhn has been much criticized for the ambiguity of his notion of paradigm. One of the most famous criticisms of the multiple meanings of the term is an often-quoted paper by Masterman in which she identifies no less than twenty-one different senses of the term in *Structure* (Masterman 1970, p. 61). However, many of these merely elucidate one another, and Masterman therefore grouped the different senses of the term into three main groups: 1) a broad set of beliefs (which Masterman called "metaphysical paradigms" or "metaparadigms"), 2) universally recognized scientific achievements, or a set of scientific habits (which Masterman called "sociological paradigms"), and 3) concrete, classical texts (which Masterman called "artefact paradigms" or "construct paradigms").

In response to this criticism Kuhn tried to disentangle the various uses of the term in the Postscript published with the second edition of *Structure* in 1970. In the Postscript Kuhn admitted that the notion of paradigms had been used in two different senses in the first edition of *Structure*: "On the one hand, it stands for the entire constellation of beliefs, values, techniques, and so on shared by the members of a given community. On the other, it denotes one sort of element in that constellation, the concrete puzzle-solutions which, employed as models or examples, can replace explicit rules as a basis for the solution of the remaining puzzles of normal science" (p. 175). In the Postscript Kuhn therefore introduced the notion 'disciplinary matrix' to denote the entire constellation of beliefs and techniques (p. 182). A disciplinary matrix contains the symbolic generalizations, that is, the scientific laws in their most fundamental forms (for example, Newton's second law of motion, $F=ma$); beliefs about which objects and phenomena exist in the world (for example, that forces exist); values by which the quality of research can be evaluated (such as, for example, accuracy or consistency); and exemplary problems and problem solutions which he now called *exemplars* rather than paradigms.

The intention in introducing the paradigm concept was to argue for the priority of exemplars over explicit definition-like rules.[8] To Kuhn, "the determination of shared paradigms is not ... the determination of shared rules" (p. 43). Investigating the history of science one will find "the search for rules both more difficult and less satisfying than the search for paradigms" (p. 43). In attempting to identify some explicit principles and rules on which all members of a scientific community agree, the historian will face problems trying to formulate the rules explicitly. No matter how the historian tries to phrase the rules, in any given explicit formulation "they would almost certainly have been rejected by some members of the group" (p. 44). Hence, apparently scientists can "agree in their *identification* of a paradigm without agreeing on, or even attempting to produce, a full *interpretation* or *rationalization* of it. Lack of a standard interpretation of an agreed reduction to rules will not prevent a paradigm from guiding research" (p. 44).

But why is this so? To answer this question Kuhn drew on the philosopher Ludwig Wittgenstein's (1889-1951) "family resemblance" analysis of everyday concepts which showed that to use terms like "chair" or "game" unequivocally we need not know a set of characteristics that all games and only games have in common. Instead, we apply the term "game" to an activity because it bears a family resemblance to other activities that we call games.[9] Kuhn used this analysis to argue that "paradigms *could* determine normal science without the intervention of discoverable rules" (p. 46), but he was aware that this does not provide an argument that science *has* to proceed without explicit rules and definitions. This part of his argument he drew partly from an analysis of science education, partly from the phase development of science. According to Kuhn's analysis, in science education the apprentice scientists never learn concepts, laws, and theories in the abstract (cf. p. 46), but through the study of applications in which they learn to relate scientific problems by resemblance and to model them to one or another of the previous achievements (cf. p. 47). To do this they need not abstract a set of rules. History shows that this way of practicing science works well as long as the problem solutions already achieved are not questioned, that is, as long as the scientists conduct normal science (cf. p. 47). However, when prior achievements are questioned, the basis by which to judge resemblance and on which to model new achievements vanishes. In this situation "rules ... become important and the characteristic unconcern about them should vanish" (p. 47). Thus, on Kuhn's view, explicit rules and debates over rules should become important only when science enters a state of crisis. And that, he claimed, "is ... exactly what does occur" (p. 47).

3.2.3 Anomalies

The notion of 'paradigm' was introduced to characterize the consensus within a scientific community regarding both past exemplary achievements and future expectations of how to model research on these past achievements. However, these expectations of how to model future research on the past achievements may occasionally be violated. Instead, problems may turn out *not* to be solvable in consonance with the previously accepted exemplary problem solutions. Such problems resisting a solution in consonance with the expectations are what Kuhn called *anomalies*. They are "the recognition that nature has somehow violated the paradigm-induced expectations that govern normal science" (p. 52f). Anomalies may therefore question the accepted tools and understandings, suggesting that they must at least be seriously modified or even be given up if the anomalous findings are to be assimilated in such a way that they are no longer anomalous but, rather, unsurprising and expected findings.

This process by which anomalies get recognized and bring into question the accepted tools and understandings is complicated. As Kuhn emphasized, "discovering a new sort of phenomenon is necessarily a complex event, one which involves recognizing both *that* something is and *what* it is" (p. 55). Based on descriptions of the discovery of oxygen and the discovery of X-rays, Kuhn emphasized that "in both cases the perception of anomaly – of a phenomenon, that is, for which his paradigm had not readied the investigator – played an essential role in preparing the way for perception of novelty. But, again in both cases, the perception that something had gone wrong was only the prelude to discovery" (p. 57).

Since the accepted exemplary problems and problem solutions imply, among other things, ontological expectations about what exists in the world and which characteristics these objects and phenomena may have, findings that run counter to these expectations cannot be made easily. One cannot just discover a phenomenon or an object which was never anticipated to exist. If it was never anticipated, there is no category by which to classify it. At first, it cannot be perceived as a phenomenon or an object, but only as 'something wrong'. Not until a category has been formed by which it can be classified can it be discovered as an actual phenomenon or object. As Kuhn stated it: the "awareness of anomaly opens a period in which conceptual categories are adjusted until the initialy anomalous has become the anticipated" (p. 64). Thus, it is characteristic for all discoveries of new phenomena that there is a previous awareness of anomaly, and a gradual and simultaneous emergence of both observational and conceptual recognition (cf. p. 62).

However, it must be noted that not all anomalies are equally severe. Some discrepancy between theoretical predictions and experimental findings can always be found but without questioning the foundation of normal science research. As Kuhn argued: "The scientist who pauses to examine every anomaly he notes will seldom get significant work done" (p. 82). Instead, some anomalies may, at least initially, simply be neglected. For example, the prediction of Newton's gravitational law for the motion of Mercury differed slightly from the observational findings, but that did not lead anybody to seriously questioning Newton's theory on these grounds (cf. p. 81). Other anomalies may eventually find a solution in consonance with the expectations. For example, Newton's graviational law also made predictions for the motion of the Moon that differed from the experimental findings. Only after decades of research on the exact computation of the predicted value was a result obtained which was in consonance with experiments, and the anomaly vanished (cf. p. 81).

For an anomaly to be a *severe* anomaly that leads to questioning the accepted tools and understandings, the anomaly must have some special importance. Without any claims to completeness Kuhn lists three different routes by which an anomaly may gain in severity (p. 82): First, an anomaly is likely to be severe if it calls some very fundamental generalizations into question. Second, an anomaly may be severe if it calls achievements into question that have a particular practical importance. Third, the development of normal science may emphasize specific aspects of an anomaly and turn it into a severe anomaly.

But even severe anomalies are not simply falsifying instances. In this respect Kuhn differs fundamentally from Popper. According to Popper, scientific theories develop and change through a process of making bold conjectures and then attempting to falsify these hypothesis. On this view, recalcitrant problem demand the refutation of the theory. To this Kuhn disagreed: "anomalous experiences may not be identified with falsifying ones. Indeed, I doubt that the latter exist" (p. 146).[10] It is Kuhn's claim that even for severe anomalies which do cause scientists to *question* the accepted theory, the anomaly does not lead them to *abandon* it unless they have another theory to replace it. "A scientific theory is declared invalid only if an alternate candidate is available to take its place" (p. 77, similarly p. 145). Thus, in making judgements about theories, scientists do not falsify but only *compare* theories. According to Kuhn, this is a historical fact – "no process yet disclosed by the historical study of scientific development at all resembles the methodological stereotype of falsification by direct comparison with nature" (p. 77). The reason for this historical fact is that a scientific theory provides the guidelines by which research within a given area must be conducted, that is, guidelines for what are considered sensible research problems and acceptable solutions. Hence, rejecting the theory without substituting an alternative would be giving up the guidelines by which to conduct research at all. As Kuhn argued: "Once a first paradigm through which to view nature has been found, there is no such thing as research in the absence of any paradigm. To reject one paradigm without simultaneously substituting another is to reject science itself" (p. 79).

3.2.4 Crisis and extraordinary science

The above discussion of anomalies raises an obvious question: if severe anomalies cannot lead scientists to abandon a theory unless they have another theory to replace it – then where does this new theory come from? When severe anomalies occur, a pronounced professional insecurity arises; a state which Kuhn labeled 'crisis' (p. 68). This crisis in the scientific community is "generated by the persistent failure of the puzzles of normal science to come out as they should" (p. 68). The response to crises is a different kind of scientific activity than normal science; a scientific activity which Kuhn called "extraordinary research" (p. 82, 86).

In the transistion from normal science to extraordinary science, the fixation point of scientific scrutiny is gradually changed. This process starts when an anomaly ceases to be just another puzzle and develops into a severe anomaly. When this happens "more and more attention is devoted to it by more and more of the field's most eminent men. If it still continues to resist, as it usually does not, many of them may come to view its solution as *the* subject matter of their discipline. For them the field will no longer look quite the same as it had earlier" (p. 82f).

From this change in fixation point, a change in the nature of the scientific activity follows. Early attempts to solve the recalcitrant problem usually model the attempted solution quite closely on exemplary problem solutions, but as the problem continues to resist solution, the attempts to solve it may gradually diverge more and more from the solutions hitherto accepted as exemplary. As Kuhn stated it: "Through this proliferation of divergent articulations (more and more frequently they will come to be described as *ad hoc* adjustments), the rules of normal science become increasingly blurred. Though there is still a paradigm, few practitioners prove to be entirely agreed about what it is. Even formerly standard solutions of solved problems are called into question" (p. 83).[11]

The various members of the scientific community may also attack the anomalous problem differently, suggesting different minor or major adjustments to the theory, "no two of them quite alike, each partially successful, but none sufficiently so to be accepted as paradigm by the group" (p. 83). During the phase of extraordinary science the previous, firm consensus within the scientific community on the foundations of their activity may therefore be replaced by an increasing dissent. However, the willingness to develop alternatives may vary considerably for different scientists. For some scientists, the very first blurring of the rules for normal science may be sufficient to induce a new way of looking at the field (cf. p. 86). Especially, the less committed a scientist is by prior practice, the easier it is to suggest alternatives. Hence, people who are young in their field and therefore less committed to the tradition than their senior colleagues may suggest alternatives more easily. As Kuhn described them "they are men so young or so new to the crisis-ridden field that practice has committed them less deeply than most of their contemporaries to the world view and rules determined by the old paradigm" (p. 144, similarly p. 90).

Hence, a crisis starts with the loosening of the rules for normal research (cf. p. 84). In loosening the rules, a crisis may enable scientists to recognize strange results as actual anomalies rather than just that 'something is wrong', or to recognize the discovery of phenomena which the old theory did not cover or even precluded (see also chapter 3.2.3). Thus, during the state of crisis, simultaneously the rules guiding the research activity are loosened and new data in the form of the recognition of anomalies or new discoveries may be collected (cf. p. 89).

According to Kuhn, it is in the course of this development that embryonic forms of alternative theories may be developed. However, he offered only sparse descriptions of the actual processes by which the alternative is developed; resorting only to such statements as "the shape of the new paradigm is foreshadowed in the structure that extraordinary research has given to the anomaly" (p. 89), but more often such as "the new paradigm ... emerges all at once, sometimes in the middle of the night, in the mind of a man deeply immersed in crisis" (p. 90).

3.2.5 Revolution

According to Kuhn, the dissent which arises during a state of crisis may come to an end in three different ways (cf. p. 84). First, it may happen that the recalcitrant problems find a solution within the framework of the old theory. Scientists will then return to normal science research in its previous form and a consensus on the previous theory is restored. Second, the recalcitrant problems may resist solution even by the most radical new approaches. In this case scientists may finally agree that no solution is likely to be found in the present state of their field and simply set the problems aside for future scientists to solve with more developed tools, while they return to normal science in its previous form. Third, the recalcitrant problems may be solved by a new theory that gradually receives acceptance until eventually a new consensus is established among members of the scientific community on the new theory. This third possibility is what Kuhn called a 'scientific revolution'. Hence, scientific revolutions are "those non-cumulative developmental episodes in which an older paradigm is replaced in whole or in part by an incompatible new one" (p. 92).[12]

Although the notion of scientific revolutions is primarily associated with Kuhn's *Structure*, the notion has a long and complicated history that goes back several centuries.[13] The idea of pathbreaking renewal emerged during the seventeenth century, but the word 'revolution' was not used for the development of science until the eighteenth century. Influenced by major social and political upheavals such as the Glorious Revolution in Britain in 1688 and the French Revolution a century later, the term 'revolution' gradually changed its meaning from the cyclical phenomenon of recurrence which it denoted within astronomy, to the fundamental kind of change which was associated with radical political developments. In this latter sense the term revolution became attached to such episodes in the history of science as the development of mechanics during the seventeenth century, the change in chemical theory which emerged with the work of Lavoisier in the 1770s, or the change in biology which was initiated by Darwin's work in the mid-19th century. These were fundamental changes which overturned not only the reigning theories but also carried with them significant consequences outside their respective scientific disciplines. They were also rare occurrences, made possible only because a true genius had been able to transcend the usual small-step incremental development of science by taking a single, big leap instead.

The Structure of Scientific Revolutions

When Kuhn used the notion of revolutions in *Structure* he widened the notion considerably. To Kuhn, fundamental changes of theory which had a significant influence on the overall world view of both scientists and non-scientists deserved to be called scientific revolutions, but also changes of theory whose consequences remained solely within the scientific discipline in which the change had taken place.

Second, for Kuhn a scientific revolution is not an incremental development. This was in direct opposition to the logical positivist view that held the development of science to be cumulative. On this view, a new theory did not simply replace its older competitor, on the contrary, the old theory was incorporated into the new, more encompassing theory. As Nagel stated it, "the phenomenon of a relatively autonomous theory becoming absorbed by, or reduced to, some other more inclusive theory is an undeniable and recurrent feature of the history of modern science" (Nagel 1961, p. 336). For Kuhn, on the contrary, a scientific revolution "is far from a cumulative process, one achieved by an articulation or extension of the old paradigm. Rather it is a reconstruction of the field from new fundamentals, a reconstruction that changes some of the field's most elementary theoretical generalizations as well as many of its paradigm methods and applications" (p. 85). The reason why a scientific revolution cannot be a cumulative process is that the new theory is developed to solve anomalies that could not be solved within the old. Thus, the new solution to the anomaly is not just an extension of the previous theory, on the contrary, the previous theory may have precluded solutions of this sort: "if new theories are called forth to resolve anomalies in the relation of an existing theory to nature, then the successful new theory must somewhere permit predictions that are different from those derived from its predecessor. That difference could not occur if the two were logically compatible" (p. 97). Adopting a new theory in favour of the old therefore has both constructive and destructive elements. When rejecting the old theory, those of its predictions which differ from the predictions of the new theory must be rejected as well. In this sense the process has a clearly destructive element. But the new theory is only adopted if it can solve the anomalous problems on which the old theory had to give up. Thus, by adopting the new theory scientists are able to account for a wider range of natural phenomena or to account with greater precision for some of those previously known. In this sense the process also has a clearly constructive element (see also chapter 3.2.6).

Third, Kuhn claimed a scientific revolution to be a "relatively sudden and unstructured event like the gestalt switch" (p. 122).

According to Kuhn, this can be documented historically through scientists speaking of "the 'scales falling from the eyes', or of the 'lightning flash' that 'inundates' a previously obscure puzzle, enabling its components to be seen in a new way that for the first time permits its solution" (p. 122f.). On Kuhn's view, this is not just a contingent historical fact but a necessary characteristic of revolutions. His argument was that although severe anomalies may initiate a crisis, previous guidelines for doing research can only be given up when new guidelines are in place; simply rejecting the guidelines would be rejecting science itself (see chapter 3.2.3). During the crisis state large portions of experience, both anomalous and congruent, are gathered up, but the transformation of these into a new, coherent theory happens 'in a flash of intuition' (cf. p. 123, similarly p. 150).[14]

With respect to this point it is important to note a difference between the individual scientist and the scientific community as the agent experiencing a scientific revolution. This distinction is not quite clear in *Structure* and Kuhn later regretted having used the notion of a gestalt switch to characterize changes in a community because "communities do not have experiences, much less gestalt switches" and that, consequently, "to speak, as I repeatedly have, of a community's undergoing a gestalt switch is to compress an extended process of change into an instant, leaving no room for the microprocesses by which the change is achieved" (Kuhn 1989b, p. 50).

In explaining what causes the group to abandon one tradition of normal research in favor of another, Kuhn described the transfer of allegiance from one normal research tradition to another as a conversion experience (cf. p. 151). These "conversions will occur a few at a time until, after the last holdouts have died, the whole profession will again be practicing under a single, but now different paradigm" (p. 152). Hence, although each individual scientist may experience his or her conversion as a gestalt switch, for the scientific community *in toto* the transition from one normal science tradition to another may be a process substantially extended in time. So, when Kuhn stated of group conversions that "like the gestalt switch, it must occur all at once (though not necessarily in an instant) or not at all" (p. 150) the statement must be taken to mean that there are no intermediate positions to which to convert. Each scientist adopts either the one or the other of the two competing normal science traditions, but cannot gradually slide from the one to the other. Hence, until the last scientists convert to the winning tradition, the battle is a fight between two groups, each having their own tradition, and when the last scientists convert, all of a sudden the two groups reduce to one (cf. p. 152).

3.2.6 Mature science versus pre-paradigmatic science

Kuhn emphasized that the phase development of science which he described, normal science → crisis and extraordinary science → revolution → normal science, is not a pattern that arises in a science until it reaches a certain level of maturity. This developmental pattern arose at different times in history for different areas of scientific inquiry. Before a given area of inquiry reached the state of maturity required for the pattern to emerge, it consisted of a number of competing schools and subschools which disagreed on various fundamentals. Kuhn called these immature states "pre-paradigmatic" science. In the pre-paradigmatic stage scientists were not able to take a common body of belief for granted, hence, each of them "felt forced to build his field anew from its foundations" (p. 13).

Lacking a common body of belief, there are no guidelines by which to distinguish, for example, between relevant and irrelevant facts. For the pre-paradigmatic sciences "fact-gathering is a far more nearly random activity" (p. 15) than for the mature sciences, and the fact-gathering is "usually restricted to the wealth of data that lie ready to hand" (p. 15). As Kuhn maintains, "though the field's practitioners were scientists, the net result of their activity was something less than science" (p. 13).

Maturity in the form of a consensus within the scientific community on the foundation of their trade has come to various areas of scientific inquiry at different times. For physical optics maturity came at some point during the seventeenth century (p. 12f), for electricity in the first half of the eighteenth century (p. 13f). For areas such as the study of motion or statics maturity may date back to antiquity, while for the study of heredity it is a fairly recent occurrence - and, still, according to Kuhn, "it remains an open question what parts of social science have yet acquired such paradigms at all" (p. 15).

3.2.7 Incommensurability

As indicated previously, the relation between the normal science traditions separated by a scientific revolution cannot be described by incorporation or incremental growth. To describe the relation, Kuhn adopted the term "incommensurability" from mathematics.[15] Using this term, Kuhn claimed that "the normal-scientific tradition that emerges from a scientific revolution is not only incompatible but often actually *incommensurable* with that which has gone before" (p. 103 (italics added), similarly p. 112).[16]

Kuhn introduced the notion of incommensurability to cover three different aspects of the relation between the pre- and post-revolutionary normal science traditions: First, incommensurability covers a change in the set of scientific problems and in the way in which scientific problems are attacked: "the proponents of competing paradigms will often disagree about the list of problems that any candidate for [a] paradigm must resolve. Their standards or their definitions of science are not the same" (p. 148). Similarly, "scientists with different paradigms engage in different concrete laboratory manipulations" (p. 126). Second, incommensurability covers conceptual change: "within the new paradigm, old terms, concepts, and experiments fall into new relationships one with the other" (p. 149).[17] Third, incommensurability covers a change in the world of the scientists' research: the proponents of different paradigms practice their trades in different worlds" (p. 150, similarly p. 111, 121).[18]

This latter aspect is the most fundamental aspect of - incommensurability – but what exactly did Kuhn mean by "different worlds"? Elsewhere in *Structure* he stated that "though the world does not change with a change of paradigm, the scientist afterwards works in a different world" (p. 121). To make sense of these claims it is necessary to distinguish between two different senses of the term 'world': the world as the independent object which science investigates, and the world as the perceived world in which scientists practice their trade (see also chapter 5.1).

In *Structure* Kuhn argued for incommensurability as a change of world in perceptual terms.[19] Drawing on results from psychological experiments showing that subjects' perceptions of various objects were dependent on their training and experience, Kuhn suspected "that something like a paradigm is prerequisite to perception itself" (p. 113). Hence, different normal science traditions would cause scientists to perceive differently. But when discussing visual gestalt switch images, one can take recourse to the actual lines drawn on the paper. Contrary to this possibility of employing an 'external standard', Kuhn claimed that "with scientific observation ... the situation is exactly reversed. The scientist can have no recourse above or beyond what he sees with his eye and instruments" (p. 114). For Kuhn, the change in perception cannot be reduced to a change in the interpretation of stable data (cf. p. 121), simply because stable data do not exist. On this point Kuhn strongly attacked the idea of a neutral observation-language (pp.

125ff.); an attack that was launched by other scholars as well during the late 1950s and early 1960s, most notably Hanson (Hanson 1958).

These aspects of incommensurability have important consequences for the communication between proponents of competing normal science traditions and for the choice between such traditions. Recognizing different problems and adopting different standards and concepts, scientists may "talk through each other when debating the relative merits of their respective paradigms" (p. 109). But if they do not agree on the list of problems that must be solved or on what constitutes an acceptable solution (cf. p. 109, 148), there can be no point-by-point comparison of competing theories, that is, there can be no comparison "by some process like counting the number of problems solved by each" (p. 148). Instead, "when paradigms enter, as they must, into a debate about paradigm choice, their role is necessarily circular. Each group uses its own paradigm to argue in that paradigm's defense" (p. 94). On this view, theory choice is a conversion that cannot be "forced by logic and neutral experience" (p. 150).

This view has led many critics of Kuhn to the misunderstanding that he saw theory choice as devoid of rational elements (see chapter 5.3). However, Kuhn did emphasize that although theory choice cannot be justified by proof, this "is not to say that no arguments are relevant or that scientists cannot be persuaded to change their minds" (p. 152). According to Kuhn, such arguments are, first of all, about whether the new theory can solve the problems that have led the old theory to a crisis (cf. p. 153), whether the new theory displays a quantitative precision strikingly better than its older competitor (cf. p. 153f), and whether the new theory predicts phenomena that had been entirely unsuspected while the old one prevailed (cf. p. 154). These arguments "are ordinarily the most significant and persuasive" (p. 155), but apart from such arguments based on the comparative ability to solve problems, aesthetic arguments may enter as well. Hence, scientists may also reject an old theory in favor of a new theory that seem "neater", "more suitable", or "simpler" (cf. p. 155). Another common misunderstanding of Kuhn's notion of incommensurability is that it is taken to imply a total discontinuity between the normal science traditions separated by a scientific revolution. However, Kuhn emphasized that "since the new paradigms are born from old ones, they ordinarily incorporate much of the vocabulary and apparatus, both conceptual and manipulative, that the traditional paradigm had previously employed" (p. 149). In this way, parts of the achievements of a normal science tradition will always prove to be permanent, even through a revolution (cf. p. 25). Consequently, "though new paradigms seldom or never possess all the capabilities of their predecessors, they usually preserve a great deal of the most concrete parts of past achievement and they always permit additional concrete problem-solutions besides" (p. 169). On Kuhn's view, incommensurability is a relation that holds only between minor parts of the object domains of two competing theories.[20]

4
Scientific Concepts

Starting with the Lowell Lectures held when he was only twenty-nine, Kuhn struggled to sketch, and progressively fill in, a consistent picture of science, the nature of scientific knowledge, and the nature of scientific concepts. Thus, some of the most important themes in Kuhn's work after *The Structure of Scientific Revolutions* were to refine and substantiate the claims advanced in *Structure* on concepts and conceptual structures.

One of the vehicles by which he developed his theory of concepts was his early reflections upon science teaching and its role in furthering and sustaining a consensus within the scientific community (chapter 4.1). His views on science teaching developed from the observation that it is an education based entirely on prepared teaching materials which the student is not supposed to question, and that it therefore serves as a rigorous training in convergent thought (chapter 4.1.1). Further, building almost exclusively on exemplary problems and concrete solutions rather than on abstract descriptions and definitions, this teaching confers the ability to recognize resemblances between novel problems and problems that have been solved before (chapter 4.1.2)

Starting from this work on exemplars as the vehicle of science teaching, Kuhn gradually developed a theory of concepts based on family resemblance (chapter 4.2). Illustrating his theory with an example of how a child might learn to recognize waterfowl, Kuhn gradually developed a full-fledged family resemblance theory of concepts (chapter 4.2.1). Apart from illustrating his theory with examples of everyday concepts, Kuhn intended to make an additional test of the plausibility of his theory by computer simulations. He was never able to carry out the simulations, but the general approach was later shown viable by other scholars with the development of neural nets (chapter 4.2.2).

In developing a family resemblance account of theories, Kuhn drew on the work of the later Wittgenstein. Although Wittgenstein's work had been subject to severe criticism, Kuhn believed that his own account of family resemblance would solve some of the philosophical problems which Wittgenstein's account had faced (chapter 4.3).

Scientific Concepts

On the basis of his theory of concepts, Kuhn rephrased *Structure's* insights about revolutions and incommensurability as claims on conceptual change and untranslatability (chapter 4.4); claims that were inspired by but also critical of the work of the philosopher Willard Van Orman Quine (chapter 4.4.1). The analysis of conceptual changes enabled Kuhn to start analyzing the conceptual microprocesses that happen during a scientific revolution (chapter 4.4.2). Through emphasizing the taxonomic structure of conceptual systems based on family resemblance, the key mechanism of scientific change, resolution of anomalies, could be rephrased as resolving violations of the principle of no-overlap between concepts that has to hold for a consistent taxonomy (chapter 4.4.3).

4.1 The role of science teaching

In developing his views on the nature of science and scientific development, very early on Kuhn started drawing on observations on the nature of science education. Thus, in the first presentation of his developing philosophical views since the 1951 Lowell Lectures he turned to an investigation of science education rather than to historical examples in substantiating the claims he advanced on the nature of science and its development.[1] This paper – 'The Essential Tension: Tradition and Innovation in Scientific Research' (Kuhn 1959a) – was delivered at the Third University of Utah Research Conference on the Identification of Scientific Talent in 1959. Another paper on this topic, 'The Function of Dogma in Scientific Research', was delivered two years later at the conference 'Scientific change: Historical studies in the intellectual, social and technical conditions for scientific discovery and technical invention, from antiquity to the present' (Kuhn 1963a).

The link between scientific practice and science education is the consensus that characterizes normal scientific research into which new members of the profession have to be brought. Thus, Kuhn claimed that "normal research, even the best of it, is a highly convergent activity based firmly upon a settled consensus acquired from scientific education and reinforced by subsequent life in the profession" (Kuhn 1959a/1977a, p. 227). This meant that a certain type of science education, one that furthered and sustained consensus, was a necessary prerequisite for scientific advancement. His argument was built, first, on the observation that science education is a rigorous training in convergent thought, and second, on the historical observations that this kind of science education had been developed at various times in history for different scientific disciplines and that science had progressed much faster afterwards. Thus, Kuhn claimed that "a rigorous training in convergent thought has been intrinsic to the sciences almost from their origin" and that the sciences "could not have achieved their present state or status without it" (Kuhn 1959a/1977a, p. 228).

4.1.1 The aim of science education

The rigorous training in convergent thought which Kuhn saw science education to be is an education based entirely on textbooks. Students are neither introduced to the history of their discipline, nor are they supposed to read original research contributions until late in their graduate years when they start conducting research of their own. Only the currently accepted results are considered relevant for the novice practitioners of the profession, whereas discarded views and theories can be ignored, as can the path that lead to the discovery of the accepted results. Further, in contrast to textbooks within the humanities or the social sciences, science textbooks covering the same field do not exemplify different approaches to their field, but differ only in level or detail. Obviously, there is an agreement within the scientific community about what it is that every student in the field must know. Hence, science education is dogmatic in the sense that students are taught only what is considered to be the correct view – but neither its history, nor its alternatives. Instead, the currently accepted view is presented as the only acceptable view. It is by this rigorous training through the exclusive use of textbooks that science education promotes convergent thought.

Kuhn's emphasis on convergent thought was a provoking view, not only among philosophers but also among science educators. World War II with its atomic bombs, radars and similar advanced scientific developments had shown the immense importance of science and scientific research. The Cold War reinforced this view. In 1950 the National Science Foundation had been formed. Interested in improving selection procedures to identify the young researchers who had the greatest potential for making important contributions to science, the NSF started supporting research on creativity, including the Utah Conferences on the Identification of Creative Scientific Talent. Kuhn's ideas on science education and convergent thought were first advanced at the third of these conferences held in 1959, but contrary to most other participants Kuhn explicitly spoke against the emphasis on creativity, wondering instead "whether flexibility and open-mindedness have not been too exclusively emphasized as the characteristics requisite for basic research" (Kuhn 1959a/1977a, p. 226). Presented with a revised version of Kuhn's argument two years later (Kuhn 1963a), B. Glass, chairman of one of the science curriculum studies supported by the NSF, declared that he was "appalled to think that, if Mr. Kuhn is right, we should go back to teaching paradigms and dogmas" (Kuhn et al. 1963, p. 382). Later, other science educators and educational philosophers, most notably Siegel (e.g. Siegel 1978), have also argued against Kuhn's account of science education, mainly because his views were seen as incompatible with the educational ideal of critical thought.

However, it is important to note that according to Kuhn, the research activity based on a settled consensus acquired from scientific education is not only immensely effective in solving the problems which the reigning traditions defines, it also serves well to identify and call attention to the loci of trouble. According to Kuhn, the continuing attempt to elucidate a currently received tradition has again and again revealed unsolvable problems and thus has eventually led to the recognition that something was fundamentally wrong and had to be changed. In this way, "the ultimate effect of [the] tradition bound work has invariably been to change the tradition" (Kuhn 1959a/1977a, p. 234).[2]

This apparent paradox was what Kuhn called 'the essential tension', that is, the tension implicit in scientific research between the convergent activity of normal science and the divergent activity of scientific revolutions. For Kuhn, convergent and divergent thinking were evidently equally important for the development of science, hence, science education had to both convey the convergent thought of the reigning paradigm, and provide the resources for a later divergence to arise.

4.1.2 The means of science education

It still remains to be explained what the consensus within the scientific community is actually about and how it is conveyed to the novices. In answering this question Kuhn observed that the established view is conveyed to the students through *exemplary problems* and their *concrete solutions*. Textbooks do not describe the sort of problems that the discipline deals with in the abstract, but "exhibit concrete problem solutions that the profession has come to accept as paradigms, and they

Scientific Concepts

ask the student, either with a pencil and paper or in the laboratory, to solve for himself problems very closely related in both method and substance to those through which the textbook or the accompanying lecture has led him". (Kuhn 1959a/1977a, p. 229). Thus, the term 'paradigm' first entered Kuhn's work to denote standard scientific problems.[3]

During the subsequent development of his position Kuhn elaborated on the view that exemplars are the vehicle by means of which new members of the scientific community acquire the ability to practice their trade. In its initial stage of development, the argument was based on the scientist's ability to identify and solve scientific problems without taking recourse to abstract rules. Correspondingly, the argument had been developed from observations on science education and seemed initially to concern only the novice. Turning to the expert one might think that although scientists once had learned to identify scientific problems by resemblance to exemplars rather than by rules, they might well later have abstracted rules for themselves. However, Kuhn found little reason to believe this. His scepticism was grounded in two observations. First, scientists "are little better than laymen at characterizing the established bases of their field, its legitimate problems and methods" (Kuhn 1970a, p. 47). Second, if one studies the history of scientific research one will note "the severe difficulty of discovering the rules that have guided particular normal-science traditions" (Kuhn 1970a, p. 46). Thus, the claim was developed into a general claim about scientific practice *in toto* when Kuhn maintained that what the research problems within a given discipline "have in common is not that they satisfy some explicit or even some fully discoverable set of rules and assumptions that gives the tradition its character and hold upon the scientific mind. Instead, they may relate by resemblance and by modeling to one or another part of the scientific corpus which the community in question already recognized as among its established achievements" (Kuhn 1970a, p. 45f).

In the first edition of *Structure,* Kuhn referred briefly to Wittgenstein in vindicating this claim. Wittgenstein had earlier advanced a view on everyday concepts which dismissed rules in favor of 'family resemblance'. From a thesis on the identification of scientific problems, Kuhn's work was growing into a family resemblance account of concepts generally.

4.2 A family resemblance account of concepts

Traces of the discussion whether concepts are defined by necessary and sufficient conditions which hold for all instances of the individual concept, or if a given concept can only be explicated by typical examples can be found as early as Plato's dialogues.[4] In modern philosophy, the predominant view since Frege has been that concepts can be defined by a set of characteristics which are individually necessary and jointly sufficient for an object to be an instance of the defined concept.[5]

However, in his *Philosophical Investigations* published posthumously in 1953, Wittgenstein launched an attack on this view. Examining the concept 'game' Wittgenstein showed that it might be impossible to find a definition of the concept which would apply to all its instances. Thus, it might well be the case that one instance of a concept shares some characteristics with another, while this latter instance shares different characteristics with a third instance of the same concept. Although in this case it is true that all instances share some characteristics with each of their fellows, no single characteristic or set of characteristics has to be shared by all instances. Not all games have sides, or a winner and a loser. Board games and political games may all rely on strategies, but differ in the level of seriousness and entertainment value. But if no single feature or set of features has to be shared by all instances it becomes impossible to define a concepts by a list of individually necessary and jointly sufficient characteristics. Instead, Wittgentein claimed, instances of a concept might bear no more than a family resemblance to each other, that is, a complicated network of overlapping and crisscrossing relations (Wittgenstein 1953, §66).

Drawing originally on this notion of family resemblance, Kuhn advanced an account of concepts based on similarity rather than rules; an account that was developed gradually over the last three decades of his life (Kuhn 1970a, 1970b/1977a, 1970c, 1974/1977a, 1979b, 1983a, 1983b, 1989a, 1990, 1991a, 1993).

According to this account, a conceptual structure is constituted by a grouping in similarity classes of the objects referred to by the concepts, for example a grouping of waterfowl into the similarity classes of ducks, geese and swans. The grouping is not determined by necessary and sufficient conditions, on the contrary, the objects in a similarity class bear no more than a family resemblance to each other. Ducks are not grouped together as ducks because they satisfy an explicit list of criteria which are individually necessary and jointly

Scientific Concepts

sufficient for something to be a duck, but because they resemble each other. Hence, there are no restrictions on *which* characteristics can be used when judging objects similar or dissimilar. As Kuhn states it: "in matching terms with their referents, one may legitimately make use of anything one knows or believes about those referents" (Kuhn 1983a, p. 681).

Denying any restrictions on which characteristics must be used in classifying objects, Kuhn claimed that it makes no sense to distinguish between essential and accidental properties. Thus, in a series of papers Kuhn argued explicitly against the assumption of essential properties, for example in the causal theory of reference which became popular among Anglo-Saxon philosophers during the 1980s (see chapter 5.1.3). Likewise, as he denied any restrictions on which characteristics to use in classification, Kuhn also claimed that it makes no sense to distinguish between characteristics that are used to classify objects and characteristics that say something about how instances of a given category behave. One may classify ducks by their color and the length of their neck and know about ducks that they have a particular body shape. But one may also classify ducks by their body shape and beak and know about ducks that they have a particular color or neck. Hence, according to Kuhn it does not make sense to distinguish between learning about concepts and learning about the world (cf. Kuhn 1964/1977a, p. 258).[6]

Positing a family resemblance account of concepts in direct opposition to the reigning view, Kuhn had to prove the plausibility of his account. Thus, he had to show that exemplar based learning is actually possible, that is, that students would acquire the ability to identify research problems of their discipline from family resemblance to other problems, without necessarily being told in what respect the problems should resemble each other.

4.2.1 Concept acquisition

According to Kuhn's account, one learns a concept by being guided through a series of encounters with objects that highlight the relations of similarity and dissimilarity currently accepted by a particular community of concept users. Hence, teaching and learning depends upon examining similar or dissimilar features of a range of objects (Kuhn 1974/1977a, 1979). However, for advanced concepts involved in scientific research problems this process of category transmission is "excessively complex" (Kuhn 1974/1977a, p. 309). To simplify matters Kuhn therefore developed an example of the transmission of a far more simple concept: a child learning to distinguish waterfowl (Kuhn 1974/1977a).[7]

In this example, the child Johnny learns to distinguish ducks, geese and swans by being guided through a series of ostensive acts by an adult familiar with the classification of waterfowl. Thus, the child is shown various instances of all three concepts, being told for each instance whether it is a duck, a goose, or a swan. Also, the child is encouraged to try to point out instances of the concepts. At the beginning of this process the child will make mistakes, for example mistaking a goose for a swan. In such cases the child will be told the correct concept to apply to the instance pointed out. In other cases the child ascribes the instance pointed out to the correct concept, and gains praise. After a number of these encounters the child has acquired the ability to identify ducks, geese, and swans as competently as the person instructing him or her.

During the ostensive acts the child encounters a series of instances of the various waterfowl, and these instances are examined in order to find features with respect to which they are similar or dissimilar. In this learning process, "the primary pedagogic tool is ostension. Phrases like 'all swans are white' may play a role, but they need not" (Kuhn 1974/1977a, p. 309). It is in this way that a conceptual structure is established by a grouping of objects into similarity classes corresponding to the extension of concepts. As we have seen from the example of the child, the grouping can be achieved solely by learning to identify similarities between objects within a particular similarity class and dissimilarities to objects ascribed to other similarity classes. Hence, for simple categories like 'duck', 'goose', and 'swan', categories are transmitted from one generation to the next solely by extracting similarity relations from the exemplars on exhibit.

Based on this detailed example of the acquisition of the concepts 'duck', 'goose', and 'swan', Kuhn claimed that "the same technique, if in a less pure form, is essential to the more abstract sciences as well" (Kuhn 1974/1977a, p. 313).[8] But rather than learning to apply concepts to objects, what one learns in the abstract sciences by this technique is to apply laws – or, in terms of his disciplinary matrix (see chapter 4.2.2), symbolic generalizations – to problem situations. However, a law should be understood here rather as a law-*sketch* or law *schema* – its detailed expression varies for different applications. For example, the law-sketch F=ma, Newton's second law of motion, applies to the problem of free fall in the form $mg=md^2s/dt^2$, to the problem of the simple pendulum in the form $mg \cdot \sin\theta = -md^2s/dt^2$, and to more complex situation in forms that makes resemblance to the law-sketch F=ma more difficult to discover.

Thus, in learning scientific concepts the student is presented with a variety of problems which can be described by various expressions of a symbolic generalization. In this process, the student discovers a way to see each problem as *like* a previously encountered problem. Recognizing the resemblance, the student "can interrelate symbols and attach them to nature in the ways that have proved effective before. The law sketch, say f=ma, has functioned as a tool, informing the student what similarities to look for, signaling the gestalt in which the situation is to be seen" (Kuhn 1970a, p. 189). Hence, Kuhn claimed that, in principle, advanced scientific concepts are acquired by the same similarity-based process as everyday concepts such as 'duck', 'goose', and 'swan'. Instead of being presented with various waterfowl, being told whether they are ducks, geese, or swans, students are presented with various problem situations and are at first shown the appropriate expression of a law sketch by which the problem can be solved. Next, the students are presented with further problem situations and must try to assign the appropriate expression for themselves. In this process, the students examine the problems in order to find features with respect to which they are similar or dissimilar. Thus, a conceptual structure is established by grouping problem situations into similarity classes corresponding to the various expressions of the law sketch.[9] As Kuhn stated it: "The resultant ability to see a variety of situations as like each other ... is, I think, the main thing a student acquires by doing exemplary problems..." (Kuhn 1970a, p.189).

4.2.2 Empirical vindication by computer simulation

As a further proof of the viability of this transmission procedure, Kuhn started developing a computer program that modeled the transmission of categories from expert to novice. The idea was that a stimulus, in the form of a string of n ordered digits, would be given as input to the machine. The machine would then transform this input into another datum representing the appropriate concept to apply to the given stimulus by the application of a preselected transformation to each of the n digits. Kuhn never finished the program, but his ideas bear much resemblance to supervised learning in neural nets.

His work on the computer simulation was carried out exclusively during the late 1960s, and traces of it are found in the three papers 'Contribution to the discussion of New Trends in History' (Kuhn et al.

1969), 'Reflections on my Critics' (Kuhn 1970c) and 'Second Thoughts on Paradigms' (Kuhn 1974/1977a), and in the postscript to the second edition of *Structure* (1970a). Only in 'Second Thoughts' and in 'Reflections' can a very brief description of the design be found. Apparently, what he attempted to design was a program which could develop the transformation function necessary to transform stimuli into clusters of data, given some previous sets of stimuli and information about which stimuli must be transformed to be placed in the same clusters and which in different (Kuhn 1970c, p. 274; Kuhn 1974/1977a, p. 310).

The approach was built on an idea of perceptual space as an n-dimensional quality space in which the distance between data, measured with a suitable metric, represents the similarity between these data. Kuhn then argued that if the various data form clusters in this space, representing different categories, it should be possible to create a program that transforms multi-dimensional input data (stimuli) to an output data representing the category to which the stimuli belongs. Further, it should be possible to create a program that designs the appropriate set of transformation functions for itself from an initial series of stimuli together with information on which stimuli must be placed in the same and which in different clusters (cf. Kuhn 1974/1977a, p. 310).

The strategy which Kuhn here described is not unlike that of supervised learning in neural nets, however, Kuhn had no contact with contemporary researchers working on this topic within the framework of AI. By the 1950s, several methods to extract functions that would transform multi-dimensional inputs into specified outputs had been developed, most notably Rosenblatt's perceptron (Rosenblatt 1962). However, in 1969 Minsky and Papert showed that the abilities of the one-layer neural net of the perceptron was very limited (Minsky & Papert 1969). For example, the perceptron was unable to model a simple logical function like exclusive-or. The criticism of Minsky and Papert was later seen as the turning point leading the majority of the AI researchers into logic-based AI. Hence, most research efforts were directed towards the development of rule-based programs – exactly the kind of approach which Kuhn strongly opposed.[10]

Kuhn's own program was never developed beyond techniques for building a quasi-random collection of clusters in an n-dimensional space, that is, the clusters which the next state of the program would teach another machine to identify. Simultaneously, the research in neural nets which might have provided Kuhn with the solution was arriving at a deadlock. Later, after neural net research had gained renewed success during the 1980s, computationally inclined philosophers like Paul Churchland (1942-) developed arguments on non rule-based concept learning which clearly resemble Kuhn's idea (cf. Churchland 1989, 1990, 1992).

4.3 Philosophical problems

Scepticism against the family resemblance account of concepts has been widespread among philosophers ever since it was first advanced by Wittgenstein. Critics of Wittgenstein's work have argued from the outset that since we can always find *some* resemblance between instances of one concept and those of another, family resemblance does not suffice to limit the extension of concepts.[11]

Kuhn was aware of this problem, but rejected taking recourse to an explicit specification of the properties that are relevant to similarity: "It is a truism that anything is similar to, and also different from, anything else. It depends, we usually say, on the criteria. To the man who speaks of similarity or of analogy, we therefore at once pose the question: similar with respect to what? In this case, however, that is just the question that must not be asked" (Kuhn 1974/1977a, p. 307; similarly Kuhn 1970a, p. 200).

Instead, Kuhn attempted to solve this problem by including not only similarity between members of the same class, but also dissimilarity to members of other classes. There is no doubt that Kuhn considered this inclusion of dissimilarity to be an important contribution to the discussion of family resemblance. Thus, in a relatively early paper he explicitly stated that one should "note that what I have here been calling a similarity relation depends not only on likeness to other members of the same class but also on difference from the members of other classes. ... Failure to notice that the similarity relation appropriate to determination of membership in natural families must be triadic rather than diadic has, I believe, created some unnecessary philosophical problems which I hope to discuss at a later date" (Kuhn 1976b, p. 199). Despite his intention, Kuhn never published a full and explicit version of his argument, and the following is a reconstruction drawn from remarks scattered over several of his papers.[12]

In talking about family resemblance Kuhn focused on the dissimilarity between concepts in a *contrast set,* which is a set of concepts which are all subordinates to the same superordinate concept (cf. Kuhn 1983a, p. 682; 1991a, p. 4; 1993, pp. 317ff). For example, the concepts 'duck', 'goose', and 'swan' are all subordinates to the superordinate concept 'waterfowl'.[13] This may seem as if Kuhn was simply taking recourse to definitions per genus and difference (or, definition *per genus proximum et differentiam specificam*), that is, a definition which first identifies the larger class (genus proximus) of which the definiendum is a subclass, and then identifies the characteristics (differentiae specificae) that distinguishes the members of that subclass from members of all other subclasses of that genus. However, this is not the case; the subdivision is not defined by particular *differentiae specificae*, but by patterns of dissimilarity which may be overlapping and criss-crossing, and which may differ completely for different speakers.

Since they are all subordinates to the same superordinate concept, the contrasting concepts together form a family resemblance concept at this superordinate level, and their instances can therefore be assumed to be more similar to each other than to instances of concepts outside the contrast set. For example, ducks, geese, and swans together form a family resemblance category of waterfowl whose members resemble each other more than they resemble members of contrasting categories such as songbirds. To avoid the problem that instances of different but highly similar categories might be mistaken for each other, Kuhn emphasized the need in concept acquisition for learning contrasting concepts together: "establishing the referent of a natural-kind term requires exposure not only to varied members of that kind but also to members of others – to individuals, that is, to which the term might otherwise have been mistakenly applied" (Kuhn 1979b, p. 413). Obviously, this analysis can be extended to new superordinate and subordinate levels. Just as the superordinate concept 'waterfowl' can be divided into the contrasting subordinates 'duck', 'goose' and 'swan', so too each of the subordinate concepts can be further subdivided into the particular species of ducks or geese or swans. The hierarchical conceptual structure that arises is one in which a general category decomposes into more specific categories that may again decompose into yet more specific categories, and so forth, in other words a taxonomy. Drawing on the dissimilarity between members of contrasting concepts, family resemblance therefore gets tied to taxonomies. Kuhn never stated this argument explicitly, but only noted that "a fuller discussion of resemblance between members of a natural family would have to allow for hierarchies of natural families with resemblance relations between families at the higher level" (Kuhn 1970b, p. 17, fn. 1).

Still, it needs to be explained how the use of contrast sets may solve the problem facing the notion of family resemblance, namely, that in principle, anything is similar to anything else. To solve this problem Kuhn introduced a condition which the world must meet if a family resemblance account of concepts is to be possible. Kuhn argued that if the chains of similarity developed gradually and continuously it would be necessary to define where the extension of the one concept ends and the extension of the contrasting concept begins: "Only if the families we named overlapped and merged gradually into one another – only, that is, if there were no *natural* families – would our success in identifying and naming provide evidence for a set of common characteristics corresponding to each of the class names we employ" (Kuhn 1970a, p. 45). On the contrary, if the chains of similarity do not develop gradually and continuously, the contrasting concepts can be used to mutually limit each others' extensions. The extension of the concept 'duck' stops with those instances that resemble geese or swans *more* than ducks. This works only if there are no intermediate cases that resemble ducks and, say, geese equally. Hence, the possibility of classifying objects into family resemblance classes depends on an "empty perceptual space between the families to be discriminated" (Kuhn 1970a, p. 197, fn. 14. Similarly Kuhn et al. 1974, pp. 508f.). On this point too, Kuhn explicitly claimed to have moved beyond Wittgenstein: "Wittgenstein ... says almost nothing about the sort of world necessary to support the naming procedure he outlines" (Kuhn 1970a, p. 45, fn. 2).

4.4 Conceptual revolutions

On the basis of his account of concepts, from the late 1960s onwards Kuhn narrowed the notion of incommensurability to cover translation failure between pairs of conceptual structures (4.4.1). By the same token, he identified revolutionary developments as those that demand changes in the conceptual structure, in contrast to developments that imply only additions or refinements to an existing structure. On this background he began analyzing the conceptual microprocesses by which changes in the conceptual structure may come about (chapter 4.4.2), and the principles which all conceptual structures have to obey (chapter 4.4.3).

4.4.1 Kuhn and Quine on (un)translatability

In *Structure* Kuhn had characterized incommensurability as a difference in the set of problems and standards for problem solutions, in ontological commitments, and in meaning. With this characterization incommensurability served to explain the problems involved in theory comparison during a scientific revolution. Focusing on the role of concepts and conceptual structures, the problem of theory comparison was rephrased as the problem that "the point-by-point comparison of two successive theories demands a language into which at least the empirical consequences of both can be translated without loss or change" (Kuhn 1970c, p. 266, similarly 1983a, p. 670).

This emphasis on translation related the work of Kuhn to Quine's work on the indeterminacy of translation. In his 1960 *Word and Object* Quine (1908-) analyzed how one may translate utterances of an unknown language on the basis of verbal response to perceptual stimulations. Quine claimed that for so-called 'radical translations', that is, translations between unrelated languages that have evolved in unrelated cultures, it may be possible to find several different translations which are logically incompatible but nevertheless all empirically equivalent. For example, when a Quinean translator meets a foreigner pointing to a rabbit while uttering 'gavagai', the translator may suggest translating the utterance as 'rabbit', or as 'brief temporal segments of rabbits', or as 'undetached rabbit-parts'. According to Quine, these different suggestions are logically incompatible, yet empirically equivalent when judged on the basis of overt behavior in observable situations. Quine's claim that it is always possible to find several logically incompatible but empirically equivalent translations has become known as the indeterminacy of translation thesis.

Kuhn critically discussed Quine's thesis in several papers (Kuhn 1970c, 1983a, 1989a, 1990a). Kuhn first adopted Quine's thesis in support of his own claim that "proponents of different theories ... speak different languages – languages expressing different cognitive commitments, suitable for different worlds. Their abilities to grasp each other's viewpoints are therefore inevitably limited by the imperfections of the processes of translation and of reference determination" (Kuhn 1977a, p. xxii, similarly 1970c, p. 268f.). Kuhn used this point to support his claim that theories expressed in different 'languages', that is, incommensurable theories, cannot be compared on their empirical implications point-by-point, arguing that according to Quine's thesis a statements from the one theory can be translated into the other theory in multiple, incompatible ways, thus questioning the outcome of a point-by-point comparison.

Later, Kuhn introduced a distinction between translation from one lanugage to another which are both known to the translator and interpretation of a language initially unknown to the interpreter. On the basis of this distinction Kuhn claimed that a Quinean translator is in fact an interpreter who *learns* a new language (cf. Kuhn 1983a, pp. 672f.). But Kuhn emphasized that "acquiring a new language is not the same as translating from it into one's own. Success with the first does not imply success with the second" (Kuhn 1983a, p. 673, similarly 1989a, p. 11; 1990, p. 300; 1991a, p. 4). He used this distinction to criticize Quine's indeterminacy of translation thesis and develop instead a more radical untranslatability thesis. Quine had argued that when suggesting different translations, or, as Quine termed them, different 'analytical hypotheses', the point is not that we cannot be sure which one is the correct one, but that there is not even an objective matter to be right or wrong about (cf. Quine 1960, pp. 73ff.). In his first discussion of Quine's 'gavagai' example, Kuhn accepted that there might be several incompatible translations. Kuhn added that "evidence relevant to choice among these alternatives will emerge from further investigation, and the result will be a reasonable analytic hypothesis with implications for the translation of other terms as well. But it will only be a hypothesis ...; the result of any error may be later difficulties in communication; when it occurs, it will be far from clear whether the problem is with translation and, if so, where the root difficulty lies" (Kuhn 1970c, p. 268). With this addition Kuhn started questioning Quine's claim that there could be no empirically detectable difference between the different translations. However, it must be noted that the empirical differences which on Kuhn's view may be detected between different translations seem to go beyond the mere behavioral response to given perceptual stimulations and thus go beyond what Quine would accept as empirical evidence.

Later Kuhn's criticism of Quine's claim became explicit when he declared that "there need be not English description coreferential with the native term 'gavagai'. In learning to recognize gavagais, the interpreter may have learned to recognize distinguishing features unknown to English speakers and for which English supplies no descriptive terminology. Perhaps, that is, the natives structure the animal world differently from the way English speakers do, using different discriminations in doing so. Under those circumstances, 'gavagai' remains an irreducible native term, not translatable into English" (Kuhn 1983a, p. 673). From this Kuhn concluded that rather than a Quinian multitude of logically incompatible but empirically equivalent translations, often there would be no possible translations at all (cf. Kuhn 1989a, p. 11, 1990, p. 300). At the same time, Kuhn made clear that the differene between Quine's indeterminacy thesis and his own untranslatability thesis was grounded in a difference between their basic premises. Thus, while Quine had developed a theory of translation based on a purely extensional semantics, Kuhn insisted that intensional aspects had to be included as well (cf. Kuhn 1983a, p. 680) and that these intenstional aspects were to be found not in the individual concep but in the structural relations among concepts (cf. Kuhn 1983a, p. 681ff.).

4.4.2 Conceptual microprocesses

Kuhn's account of incommensurability as untranslatability is concerned with the difference between two fully developed conceptual structures (cf. Kuhn 1983a, p. 672). However, the account raises the interesting question how a new conceptual structure develops from the old, that is, how conceptual changes come about.

By the gestalt switch metaphor he had introduced in *Structure*, Kuhn seemed to be indicating that conceptual changes happen all at once. However, during the 1980s he started admitting that these interpretations of revolutions and later of translation failure were both the result of his experiences as a historian discovering the past, and that the experience of the scientists moving through time in the opposite direction may be different. Admittedly, he still argued that scientists experience gestalt switches, but he now added the qualification that "their shifts in gestalt will ordinarily be smaller than the historian's for what the latter experiences as a single revolutionary change will usually have been spread over a number of such changes during the development of the sciences" (Kuhn 1983b, p. 715).

However, he still denied that these language changes might have taken place originally as what he called 'gradual linguistic drift'. In support of his view he referred to empirical evidence in the form of reports of 'aha' experiences (Kuhn 1983b, p. 715). Further, Kuhn argued that "it is the acceptance of fuzziness that permits drift, the gradual warping of the meanings of a set of interrelated terms over time" and that "in the sciences borderline cases of this sort are sources of crisis", and he concluded that although gradual linguistic drift may happen in discourses such as those of the political life, it is simply inhibited in the sciences (Kuhn 1983b, p. 715).

By the end of the 1980s, Kuhn returned to a discussion of the difference between conceptual differences as experienced by the historian working backwards in time and the conceptual changes created by the scientists taking part in the scientific development. He still adhered to the view that historians, working backwards, may experience only a single conceptual shift whereas the actual developmental process required a whole series of stages. However, he now emphasized that a conceptual structure is a property constitutive of a scientific *community* and also a property which is carried by each individual member of the community. One must therefore note that "as the conceptual vocabulary of a community changes, its members may undergo gestalt switches, but only some of them do and not all at the same time... To speak ... of a community's undergoing a gestalt switch is to compress an extended process of change into an instant, leaving no room for the microprocesses by which the change is achieved" (Kuhn 1989b, p. 50).

In addressing the question how a new conceptual structure develops from the old, two aspects have to be distinguished: the conceptual microprocesses which occur *within the community* relating to the conceptual structure as a property constitutive of this community, and the conceptual microprocesses which occur within the conceptual structure as something carried by each individual member of the community. Kuhn never addressed the former aspect although he saw his work as "designed to leave room for their exploration" (Kuhn 1989b, p. 51). The latter aspect was only indirectly addressed as part of his work from the early 1990s on what he now called the lexical taxonomies or simply a lexicon, which are a conceptual structures covering only kind terms.

4.4.3 The scientific lexicon and anomalies as overlap

From the outset Kuhn had described conceptual structures as reflecting a grouping of objects into similarity classes (Kuhn 1970c, 1974). However, he simultaneously talked about theories as languages generally conceived, and it was not until the beginning of the 1990s that he explicitly noted that his interest was only with a restricted class of terms: "roughly speaking, they are taxonomic terms or kind terms, a widespread category that includes natural kinds, artifactual kinds, social kinds, and probably others" (Kuhn 1991a, p. 4). Hence, taxonomic terms refer to the objects and situations into which a language takes the world to be divided.[14]

Scientific Concepts

In his previous work on conceptual structures as determined by family resemblance between objects, Kuhn had briefly stated that his account 'would have to allow for hierarchies of natural families' (see chapter 4.3) but without making explicit what this would imply. Kuhn now returned to the issue of conceptual hierarchies, or taxonomies, and tried to spell out which principles the concepts in a given scientific lexicon have to obey: "No two kind terms, no two terms with the kind label, may overlap in their referents unless they are related as species to genus. There are no dogs that are also cats, no gold rings that are also silver rings, and so on; that's what makes dogs, cats, silver, and gold each a kind" (Kuhn 1991a, p. 4).

The conditions that no terms may overlap he termed the "no-overlap principle", or the "no-overlap condition". The qualification that terms may overlap in their referents if related as species to genus he only briefly mentioned as "a restriction of hierarchical relations that I cannot spell out here" (Kuhn 1991a, p. 5). Although Kuhn did not refer to it, these conditions have a long history in the philosophical literature.[15] The study of classification and the division of concepts is an important part of classical logic, and in classical treatments of logic such as, for example, Kant's *Logic*, taxonomic divisions are characterized by the three principles that the concepts formed by the division 1) do not overlap, 2) are subordinated to the same superordinate concept, and 3) together exhaust the superordinate concept (cf. Kant 1800/1992, §111).

However, Kuhn's interest was not with how a taxonomy is established by a division of the world from scratch, but with how taxonomies *change* as science develops (see also chapter 5.1). In particular, on the basis of his account of kind terms the distinction between normal science and revolutionary developments could be rephrased as a "distinction between developments which do and developments which do not require local taxonomic change" (Kuhn 1991a, p. 7). The taxonomic principles (especially the no-overlap principle which was the only one of the three which Kuhn explicitly mentioned) may therefore be seen as the key to understanding both what might provoke changes to the lexicon and what incommensurability is about.

According to Kuhn, since a taxonomy has to obey the no-overlap principle, the taxonomy will be seriously challenged if a referent turns up that violates the principle. As a response to such an occurrence the taxonomy will have to be changed: "if the members of a language community encounter a dog that's also a cat (or, more realistically, a creature like the duck-billed platypus) they cannot just enrich the set of category terms but must instead redesign a part of the taxonomy" (Kuhn 1991a, p. 4). In other words, an occurrence that violates the no-overlap principle is a severe anomaly that calls the lexicon into question and initiates extraordinary research. The extraordinary research may then result in a restructuring of the lexicon, that is, the introduction of new categories or changes in the relation among categories in such a way that the restructured lexicon comply with the no-overlap principle again.[16]

It must be noted that although a violation of the no-overlap principle calls for changes in the lexicon this does not imply that *all* of the lexicon must change. If, for example, a dog is encountered that is also a cat, changes must be made to categories of dogs and cats, but although this may entail changes to the categories of wolves and tigers, it may have no influence on the categories of cows or turtles. It must also be noted that violations of the no-overlap principle may only provide an explanation of how conceptual changes are triggered, but not how they will develop. In this respect the no-overlap principle does not alone suffice to develop a dynamics of conceptual change.[17]

In the same way that he explained how taxonomic change is triggered by the violation of the no-overlap principle, Kuhn also explained incommensurability between two fully developed taxonomies by use of the no-overlap principle. According to Kuhn, if two fully developed taxonomies contain concepts that overlap partially in their referents, such as, for example, the concepts of force in the Aristotelian and in the Newtonian taxonomies, referents in the overlap region will be subject to incompatible expectations or incompatible natural laws (cf. Kuhn 1993, p. 318f.). The force involved in projectile movements, for example, is according to the Aristotelian taxonomy a force which is exerted on the projectile by the mover and it ceases as soon as the projectile has left the mover. On the contrary, according the Newtonian taxonomy the gravitational force acts on the projectile during the whole motion.[18] In such cases, sentences articulated within the one taxonomy may be inconsistent with sentences articulated within the other.

5
Philosophical Implications

Kuhn's philosophical work from *Structure* and onwards contains controversial views on a number of philosophical issues. *Structure* soon became subject of intense scrutiny beginning with the July 1965 symposium *Criticism and the Growth of Knowledge* (Lakatos and Musgrave 1970).[1] Kuhn's notion of incommensurability has been a special focus of much discussion.

Provoked by Kuhn's repeated use of wordings like "when paradigms change, the world itself changes with them" (Kuhn 1970a, p. 111), or "the proponents of competing paradigms practice their trades in different worlds" (Kuhn 1970a, p. 121), some critics have argued that there seems to be simply no point at issue between incommensurable theories. If they are formulated in untranslatable languages, any kind of conflict between them is precluded. Hence, incommensurable theories cannot be rivals. As Shapere put it in his review of *Structure*: "if [two competing paradigms] disagree as to what the facts are, and even as to the real problems to be faced and the standards which a successful theory must meet – then what are the two paradigms disagreeing about? And why does one win?" (Shapere 1964, p. 391).

What the critics were repelled by were a number of related theses about realism, truth, and rationality. By claiming of incommensurable theories that their adherents "practice their trades in different worlds" (Kuhn 1970a, p. 121), Kuhn was questioning traditional realism (chapter 5.1.). Rejecting the view that science was zeroing in on nature's real joints or that successive theories provided successively closer approximations to nature, by the same token Kuhn had to dismiss the correspondence theory of truth (chapter 5.2). These views often led to Kuhn being accused of relativism and irrationalism. However, Kuhn maintained that, on his view, although the foundationalist account of rationality had to be rejected, it was still possible to defend an evolutionary account of rationality and thus avoid relativism. An important ingredient in this position is to view the scientific community rather than the individual scientists as the principal agent of science (chapter 5.3).

5.1. Realism

Kuhn's incommensurability thesis can be seen as a challenge to realism and it has remained an important part of the debate between realism and anti-realism that keeps raging in the philosophy of science.[2] Kuhn's non-realist stance in this debate has been an issue of interpretation, discussion and criticism from the outset. The most important line of criticism was launched originally by Scheffler who in his 1967 *Science and Subjectivity* argued that rival theories could be compared only if they share reference. This approach was later extended by other scholars, most notably Putnam. However, this referential approach to theory comparison is based on realism, and it was heavily attacked by Kuhn in a series of papers from the late 1970s and onwards (chapter 5.1.1). Kuhn's own position is not a traditional anti-realist position, but can be seen as an attempt to define a position in-between realism and constructivism (chapter 5.1.2).

5.1.1 The Kuhn-Putnam debate

The incommensurability thesis asserts that successive theories are embedded in different conceptual systems and that, as a consequence of this, a number of the terms which on a superficial view may seem to be shared by two competing theories may nevertheless differ in meaning. For example, the term 'mass' is used both within classical mechanics and within relativity theory but does not mean the same in the two contexts. For example, within classical mechanics mass is conserved, while within the relativity theory mass and energy are interconvertible (cf. Kuhn 1970a, p. 102). Likewise, competing theories may posit entities which do not exist according to its competitor. For example, prior to the chemical revolution phlogiston was taken to exist as a substance that was set free during the process of combustion from the burning body. After the chemical revolution the existence of phlogiston was denied. Instead, a new entity, oxygen, was introduced which was assumed to enter into combustion by combining with the burning body. Because of such meaning variances and ontological differences some statements of the one theory cannot be translated into statements of the other theory without residue or loss, consequently, to the extent that claims of the one theory cannot be translated into claims of the other theory, the content of these claims cannot be directly compared (see chapter 4.4).

One of the standard responses to the claim that due to translation failure incommensurable theories cannot be compared in a point-by-point manner is the referential stability approach. This approach was first proposed by Scheffler in his 1967 *Science and Subjectivity*. Scheffler argued that even if two theories are mutually untranslatable, as long as their terms share reference it is possible for statements from the two theories to conflict, hence, for the theories to be rivals and comparable.

To determine reference Scheffler adopted the classical descriptive theory of reference according to which the reference of a term is determined by its associated descriptive content. However, the descriptive theory of reference only secures referential stability in those cases in which an entity is described by different theories in different ways that are still compatible. On the contrary, if two theories offer incompatible descriptions of an entity, that is, descriptions that cannot be true of the same thing, then on the descriptive theory of reference the two theories cannot be referring to the same thing. For example, before the development of quantum mechanics, the description of particles such as the electron contained, among other things, the notion that the particle would at any given time have a particular position and a particular momentum. After the development of quantum mechanics it was ruled out that the position and the momentum of a particle could both be well-defined at the same time. Consequently, within quantum mechanics the description of a particle with both a well-defined position and a well-defined momentum refers to nothing. On this view, the term 'electron' cannot be seen as referring to the same thing before and after the development of quantum mechanics. In sum, the descriptive theory of reference was faced with the problem that a change of theory might entail new descriptions that are not true of the same things as previous descriptions. Since description determines reference such changes of description would entail change of reference, and the possibility for the theories to conflict would vanish.

Several scholars, most notably Putnam, have therefore argued for the necessity of adopting a theory of reference that would enable a term to refer although the scientists' beliefs about it, including their scientific definitions of the term, might be mistaken. On Putnam's view, an adequate theory of reference should be able to explain how scientists at the beginning of the 20th century were referring to electrons with their term 'electron' although their beliefs about electrons were mistaken, and how we are referring to the same particles even if our beliefs about electrons some day turn out to have been mistaken too.

Philosophical Implications

Putnam argued that the causal theory of reference would achieve this kind of referential stability.[3] According to the basic version of the causal theory, a term is introduced in an introducing event. At this introduction, the object or kind to which the term is to refer is singled out by ostension or by a description. In subsequent use the term continues to refer to the entity to which it was originally attached on the occasion of its introduction. In the case of kind terms, the extension of the term is fixed by means of a representative sample, and the extension of the term consists of the set of objects which bear the same-kind-as relation to objects in the original sample. The same-kind-as relation is taken to be a theoretical relation determined by the internal structural traits of the objects to which the term refers, and the details of the relation are thus to be discovered by scientific research (cf. Putnam 1975, p. 225).

For example, the term 'gold' may be introduced through an ostensive act in which a speaker points to a sample of gold and says "this is gold". In later use of the term, 'gold' will refer to stuff that bears the same-kind-as relation to the original sample. This relation is determined by the internal structural traits of gold which scientific research has revealed to be that it has the atomic number 79. Hence, on the causal theory, the term 'gold' has always referred to substances with atomic number 79, even before the periodic table of the elements was developed. Should our scientific theory change the term 'gold' will continue to refer to that natural kind under the new description.

The causal theory of reference builds on the realist assumptions that a fixed realm of "theory-independent entities" exists (Putnam 1975, p. 236) and that the aim of science is to improve the accord between our concepts and these entities. On this view terms are used "as if the associated criteria were not *necessary and sufficient conditions*, but rather *approximately* correct characterizations of some world of theory-independent entities" (Putnam 1975, p. 237, italics in the original), hence, later theories are "in general, *better* descriptions of the *same* entities that earlier theories referred to" (ibid., italics in the original).

Kuhn rejected the realist assumptions that some fixed realm of theory-independent entities exists and that our theories develop to give better descriptions of their underlying traits. Kuhn's criticism of the causal theory was primarily addressed against Putnam, and centered on his Twin Earth argument. In this argument, Putnam had imagined a Twin Earth which is exactly like our own Earth except for the single difference that what is called 'water' on Twin Earth is not the chemical compound H_2O but a different compound with a very long and complicated chemical formula, abbreviated as XYZ. This Twinearthian

Philosophical Implications

water is indistinguishable from the Earth's water at normal temperature and pressure, and it fills the oceans and lakes at Twin Earth, rains from the skies and quenches the thirst of the Twinearthians. On Putnam's view, for something to be water it has to bear the same-kind-of relation to the original sample of Earthian water, even if the details of this relation is not fully discovered by scientific research. Hence, although both Earthians and Twinearthians may originally have taken Twinearthian water to be water, with the development of chemical theory they would at some point have discovered that they only *mistook* Twinearthian water for water. Imagining a spaceship from Earth visiting Twin Earth Putnam therefore argued that although the astronauts' first supposition might probably be that the term 'water' has the same meaning on Earth and on Twin Earth, after doing some chemical analyses they would report back to Earth that on Twin Earth the word 'water' means XYZ.

Kuhn disagreed to this part of the scenario, arguing that the astronauts' report would rather be something like "Back to the drawing board! Something is badly wrong with chemical theory" (Kuhn 1989a, p. 27; 1990, p. 310). Kuhn argued that "the terms 'XYZ' and 'H_2O' are drawn from modern chemical theory, and that theory is incompatible with the existence of a substance with properties very nearly the same as water but described by an elaborate chemical formula" (Kuhn 1989a, p. 27; 1990, p. 310). Hence, to Kuhn the discovery that Twin Earthian water is not H_2O but XYZ is not merely a discovery of the underlying traits of a particular substance, but the discovery of an anomaly to chemical theory as such. Resolving that anomaly might lead to a restructured chemical lexicon, and only with this "differently structured lexicon, one shaped to describe a very different sort of world, could one, without contradiction, describe the behavior of XYZ at all, and in that lexicon 'H_2O' might no longer refer to what we call 'water'" (Kuhn 1989a, p. 27).

Further, Kuhn attacked Putnam's claim that the term 'water' refers to and has always referred to the substances with the underlying trait that its chemical formula is H_2O. Kuhn pointed out that 'H_2O' not only picks out water, but also ice and steam. The underlying trait of water is not simply 'H_2O' but 'liquid H_2O' or 'close-packed H_2O particles in rapid relative motion' (cf. Kuhn 1989a, p. 28). From this Kuhn argued that the causal theory faces problems as soon as more than one essential property is required for the same-kind-as relation: "When two non-coextensive names are required ... – 'H_2O' and 'liquidity' in the case of water – then each name, if used alone, picks out a larger class than the pair does when conjoined, and the fact that they name

Philosophical Implications

properties becomes central. For if two properties are required, why not three or four? Are we not back to the standard set of problems that causal theory was intended to resolve: which properties are essential, which accidental; which properties belong to a kind by definition, which are only contingent?" (Kuhn 1989a, p. 29).

At the core of Kuhn's criticism is his rejection of realism and the assumption that we refine our theories to give better and better descriptions of the set of entities of which the world consists. As Kuhn rhetorically asked: "Does it obviously make better sense to speak of accommodating language to the world than of accommodating the world to language? Or is the way of talking which creates that distinction itself illusory? Is what we refer to as "the world" perhaps a product of a mutual accommodation between experience and language?" (Kuhn 1979b, p. 418). Rejecting the existence of a fixed realm of theory-independent entities, suggesting instead that the world and the entities it consists of are determined both by experience and by language, Kuhn was grappling for a position which was neither purely realist, nor sheerly constructivist, but something in-between.

5.1.2 Kuhn's non-realist stance

Claiming that "though the world does not change with a change of paradigm, the scientists afterwards work in a different world" (Kuhn 1970a, p. 121), Kuhn distinguished between two different kinds of worlds: a "hypothetical fixed nature" (Kuhn 1970a, p. 118) and a "perceived world" (Kuhn 1970a, p. 128). This bears resemblance to the distinction which Kant introduced between the thing-in-itself (*Ding-an-sich*) and its appearance (*Erscheinung*), that is, the representation which it causes in us by affecting our senses (cf. Kant 1783/1950, §13, remark 2). Like Kant's thing-in-itself, Kuhn's hypothetical fixed nature is "ineffable, undescribable, undiscussible" (Kuhn 1991a, p. 12). It is a "Kantian source of stability", "located outside of space and time" (*ibid.*).[4] The perceived world, on the contrary, is "a world already perceptually and conceptually subdivided in a certain way" (Kuhn 1970a, p. 129), but this subdivision is a structure which is imposed on the world by means of the concepts applied to it.[5] Adopting Hoyningen-Huene's Kantian interpretation, Kuhn's two world concepts can be described as the "the world-in-itself" and "the phenomenal world", respectively (Hoyningen-Huene 1993, chapter 2.1).

As described in chapter 4.2, on Kuhn's theory a conceptual structure is constituted by a grouping in similarity classes of the objects referred to by the concepts. The grouping is not determined by necessary and sufficient conditions, on the contrary, the objects in a similarity class bear no more than a family resemblance to each other. Denying the existence of a fixed realm of theory-independent entities, the relations of similarity and difference that underlie the similarity classes cannot simply be discovered as the underlying traits of the fixed entities. Instead, the similarity and difference relations are constitutive of the structure of the phenomenal world, that is, of which entities exist in the phenomenal world. On Kuhn's view, there is no fixed entity of water with the underlying traits of having the chemical formula H_2O to be discovered by science. At some point in history we have divided the world into different entities on the basis of, for example, freezing point and boiling point at a specific pressure. Later we may have discovered other traits by which to distinguish these entities, but there are no restrictions on which characteristics can be used in distinguishing the entities. Some may distinguish water by freezing point and boiling point, others may distinguish it by chemical analysis.

To claim that the similarity and difference relations are constitutive of the structure of the phenomenal world may sound as if the relations of similarity and difference can be freely invented to constitute any arbitrary structure for the phenomenal world. However, this is not the case: "nature cannot be forced into an arbitrary set of conceptual boxes. On the contrary, ... the history of the developed sciences shows that nature will not indefinitely be confined in any set which scientists have constructed so far" (Kuhn 1970c, p. 263). This argument introduces some sort of 'resistance' against giving arbitrary structures to the phenomenal world. The resistance appears in the form of anomalies, that is, situations in which it becomes clear that something is wrong with the structure given to the phenomenal world by our concepts – that objects do not behave or situations do not develop as prescribed by the current conceptual structure. If we discover that something has physical characteristics just like water but chemical characteristics like a big and complicated molecule, this anomaly tells us that something is wrong with our conceptual structure. In this respect, although Kuhn's position has contructivist traits, it is not an extreme constructivist position allowing arbitrary structures of the phenomenal world.[6] On the other side, anomalies only show that a partucular structuring of the phenomenal world does not work, not how the phenomenal world has to be structured instead. If the resistance fully determined the structure of the phenomenal world there would be no need to introduce a phenomenal world different from the world that offers resistance and the position would reduce to a traditional realist

61

position. Hence, Kuhn's position it is no traditional realist position either, but can be seen as an attempt to define a position in-between realism and constructivism.

But if Kuhn rejected traditional realism, how must one then understand the premise he built on in his work on conceptual structures that there is an "empty perceptual space" between the families to be discriminated (see chapter 4.3)? This premise seems to build on realism, revealing a tension in Kuhn's position: the structure of perceptual space is determined by relations of similarity and difference between instances of contrasting concepts, but at the same time these relations are dependent on a certain structure of the perceptual space. To dissolve this tension it is important to note that the position which Kuhn was grappling for is inherently historical. On his view, the phenomenal world is never structured from scratch by its inhabitants. Instead, the inhabitants of any phenomenal world have originally found this world "already in place, its rudiments at their birth and its increasingly full actuality during their educational socialization, a socialization in which examples of the way the world is play an essential part. ... Creatures born into it must take it as they find it. They can, of course, interact with it, altering both it and themselves in the process, and the populated world thus altered is the one that will be found in place by the generations which follows" (Kuhn 1991a, p. 10). Hence, a structured perceptual space with empty space between families to be discriminated is always inherited by any generation from their predecessors.[7] But once the new generation has gained access to this particular phenomenal world they may start reshaping it by introducing new relations of similarity and difference and abandoning old ones and thus provide a different structure of the phenomenal world to their successors than the structure they inherited themselves.[8]

5.2 Truth

Rejecting the existence of a fixed realm of theory-independent entities, the goal of theory-evaluation cannot be to determine whether or not the theories correspond to an external, mind-independent world. Thus, from *Structure* and onwards Kuhn maintained that it does not make sense to talk about theories as drawing closer and closer to the truth:

"One often hears that successive theories grow ever closer to, or approximate more and more closely to, the truth. Apparently generalizations like that refer not to the puzzle-solutions and the concrete predictions derived from a theory but rather to its ontology, to the match, that is, between the entities with which the theory populates nature and what is 'really there'. Perhaps there is some other way of salvaging the notion of 'truth' for application to whole theories, but this one will not do" (Kuhn 1970a, p. 206).

The notion of truth which Kuhn was attacking is the correspondence theory of truth. According to this theory a proposition is true if it corresponds to that of which it forms a judgement. For example, the proposition "the speed of light is $2.998 \cdot 10^8$ m/s" is true if what the proposition states corresponds to reality.

This theory of truth is not the only one available at the philosophical fair, nevertheless Kuhn revealed little about what he saw as the alternative. He insisted that truth had to be lexicon-dependent: "evaluation of a statement's truth value is, in short, an activity that can be conducted only with a lexicon already in place, and its outcome depends upon that lexicon" (Kuhn 1989a, p. 24). However, this should not be taken to imply that the same proposition could in principle be assigned different truth values when embedded in different lexicons: "the point is not that laws true in one world may be false in another but that they may be ineffable, unavailable for conceptual or observational scrutiny. It is effability, not truth, that my view relativizes to worlds and practices" (Kuhn 1993, p. 336). The main quality which Kuhn wanted from a theory of truth was that it should introduce minimal laws of logic such as the law of non-contradiction (cf. Kuhn 1991a, p. 9). Hence, on Kuhn's view, truth should not be seen as an ingredient of reality but merely as serving to "require choice between acceptance and rejection of a statement or a theory in the face of evidence shared by all" (Kuhn 1991a, p. 8f.). [9]

Rejecting the view that science was drawing closer and closer to the truth, Kuhn was often accused of relativism. However, Kuhn argued that the rejection of this view does not necessarily entail relativism. Thus, Kuhn maintained that "there are shared and justifiable, though not necessarily permanent, standards that scientific communities use when choosing between theories" (Kuhn 1989a, p. 23, fn. 24. See also chapter 5.3). However, on Kuhn's view these standards can be used only to decide which of two (or more) theories is the *better* one: "judgements of this sort are necessarily comparative: which of two bodies of knowledge – the original or the proposed alternative – is *better* for doing whatever it is that scientists do" (Kuhn 1991a, p. 6). On this basis Kuhn asserted that science does progress in the sense that it shows an increasing ability for solving scientific problems:

> "Imagine ... an evolutionary tree representing the development of the scientific specialties from their common origin in, say, primitive natural philosophy. Imagine, in addition, a line drawn up that tree from the base of the trunk to the tip of some limb without doubling back on itself. Any two theories found along this line are related to each other by descent. Now consider two such theories, each chosen from a point not too near its origin. I believe it would be easy to design a set of criteria – including maximum accuracy of predictions, degree of specialization, number (but not scope) of concrete problem solutions – which would enable any observer involved with neither theory to tell which was the older, which the descendant. For me, therefore, scientific development is, like biological evolution, unidirectional and irreversible. One scientific theory is not as good as another for doing what scientists normally do. In that sense I am not a relativist" (Kuhn 1970c, p. 264).

5.3 Rationality

In *Structure* Kuhn argued that, due to incommensurability, theories cannot be compared in a point-by-point manner, but only in a circular dispute in which each group uses its own theory to argue in that theory's defense (see chapter 3.2.7). On this view, theory choice cannot be forced by logic and neutral experience: "the choice between paradigms (or theories ...) cannot be compelled by logic and experiment alone; in these matters there is no such thing as proof, no point at which the opponent of a new view violates a rule of science, begins to behave unscientifically" (Kuhn 1971a, p. 144f.).

Instead, Kuhn claimed that theory choice is a process of *conversion* which is driven by *persuasion* rather than by proof. In answering the question how such conversions are induced and resisted Kuhn admonished "what sort of answer to that question may we expect? Just because it is asked about techniques of persuasion, or about argument and counterargument in a situation in which there can be no proof, our question is a new one, demanding a sort of study that has not previously been undertaken. ... when asked about persuasion rather than proof, the question of the nature of scientific argument has no single or uniform answer" (Kuhn 1970a, p. 152).

Rejecting the traditional ideal of an algorithmic decision procedure, Kuhn's view of theory choice was met with perplexity and consternation by several critics who dismissed it as "irrational, a matter for mob psychology" (Lakatos 1970, p. 178, italics removed), or as "mere persuasive displays without deliberative substance" (Scheffler 1967, p. 81). The critics saw Kuhn as expressing the view that "the decision of a scientific group to adopt a new paradigm cannot be based on good reasons of any kind, factual or otherwise; quite the contrary, what counts as good reason is determined by the decision" (Shapere 1966/1981, p. 54). Hence, on their reading of Kuhn, "there remains *no* basis for choosing between [competing theories]. Choice must be made without any basis, arbitrarily" (Shapere 1966/1981, p. 55).

However, Kuhn never attempted to argue for irrationalism. On the contrary, he insistently rebutted: "As I have said before, both here and elsewhere, I do not for a moment believe that science is an intrinsically irrational enterprise" (Kuhn 1971a, p. 143). But whereas the traditional view among philosophers had been that if scientists diverged from the philosophical account of rational behavior this was merely a sign of human imperfection, Kuhn wanted to draw the opposite conclusion, namely that the philosophical account was in need of improvement:

> "What I have perhaps not made sufficiently clear, however, is that I take that assertion not as a matter of fact, but rather of principle. Scientific behavior, taken as a whole, is the best example we have of rationality. Our view of what it is to be rational depends in significant ways, thought of course not exclusively, on what we take to be the essential aspects of scientific behavior. That is not to say that any scientist behaves rationally at all times, or even that many behave rationally very much of the time. What it does assert is that, if history or any other empirical discipline leads us to believe that the development of science depends essentially on behavior that we have previously thought to be irrational, then we should conclude not that science is irrational but that our notion of rationality needs adjustment here and there" (Kuhn 1971a, p. 144, similarly Kuhn 1970c, p. 263f.).

Philosophical Implications

However, although arguing from history that in actual scientific practice theory choice does not seem forced by logic and neutral experience, Kuhn did not merely want to offer a descriptive account, but a prescriptive account explaining why the development of science requires a decision process that permits scientists to disagree (cf. Kuhn 1977c, p. 332).

To understand Kuhn's account of theory choice it is important to note that the principal agents of science are the scientific *communities* and not the individual scientist. Kuhn repeatedly emphasized that "our concern will not then be with the arguments that in fact convert one or another individual, but rather with the sort of community that always sooner or later re-forms into a single group" (Kuhn 1970a, p. 153); "ultimately the choice between paradigms is a community decision, that what passes for proof, verification, or falsification in the sciences has not occurred until an entire community has been converted or re-formed about a new paradigm" (Kuhn 1971a, p. 145).

The reason for this viewpoint is to be found in Kuhn's phase model of scientific development and the functional role played by individuals and communities in this development. On Kuhn's phase model, a new theory competing with the previously accepted theory is developed in response to severe anomalies that have made the reigning theory seem untenable (see chapter 3.2.3). But when anomalies are severe and when the reigning theory seems untenable are matters of judgement, and it was Kuhn's claim that a lack of unanimity is essential to the progress of science. As Kuhn argued: "Most judgements that a theory has ceased adequately to support a puzzle-solving tradition prove to be wrong. If everyone agreed in such judgements, no one would be left to show how existing theory could account for the apparent anomaly as it usually does. If, on the other hand, no one were wiling to take the risk and then seek an alternate theory, there would be none of the revolutionary transformations on which scientific development depends" (Kuhn 1970c, p. 248; similarly Kuhn 1977c, p. 332). Hence, on Kuhn's view it is a precondition for the development of new scientific theories rivaling with the old, accepted ones that different scientists can come to make different judgements of the severity of anomalies and of the quality of new theories and consequently reach different decisions as to which theory to work on.

On the other hand, divergence among scientists about which theories to pursue tend to come to an end. In the course of time, usually a new consensus concerning theory choice will be reached within the group of scientists. A process of theory choice which includes both the dissolution of the previous consensus and the emergence of a new consensus in the scientific community requires an intricate relation between individual and sociological aspects of theory choice. What needs explaining is both how the individuals can come to divergent decisions as well as how the group can come to a common decision.

To explain the possibility of consensus to first dissolve and then emerge again, Kuhn pointed out that a scientific community holds a list of values which all its members share. Among others, these values include accuracy, consistency, scope, simplicity, and fruitfulness (cf. Kuhn 1977c, p. 321f.). These values "provide *the* shared basis for theory choice", however, "individually, the criteria are imprecise; individuals may legitimately differ about their application to concrete cases" (Kuhn 1977c, p. 322). For example, a value like simplicity is not defined in any precise way, but may be interpreted differently by different scientists. Further, the individual values may "conflict with one another" (*ibid.*). For example, the most accurate and the most simple of a set of competing theories need not be the same. Hence, "when scientists must choose between competing theories, two men fully committed to the same list of criteria for choice may nevertheless reach different conclusions" (Kuhn 1977c, p. 324). On this view, when an anomaly emerges the different ways in which individual members of the scientific community employs the shared values may result in different judgements of the anomaly and different attitudes towards the various alternative theories that are developed in response to the anomaly.

The dissent continues until all members of the scientific community, on the basis of their individual interpretations and weighings of the shared values, prefer a single theory.[10] Hence, on this view, "it is the community of specialists rather than its individual members that makes the effective decision" (Kuhn 1970a, p. 200), while the temporary dissent prior to the decision serves as "an indispensable means of spreading the risk which the introduction or support of novelty always entails" (Kuhn 1977c, p. 332).

6

Later Developments

In the Preface to *Structure*, Kuhn admitted that although he had tried both to point out and to document the main philosophical implications of this book, limitations of space had made him refrain from detailed discussion of the various positions taken by his contemporaries on the same issues. Often his skepticism would be directed against a philosophical attitude rather than to one of its fully articulated expressions. As a consequence, Kuhn foresaw that some philosophers might feel that he was missing their points, and he added "I think they will be wrong, but this essay is not calculated to convince them. To attempt that would have required a far longer and very different sort of book" (Kuhn 1970a, p. x).

Since the beginning of the 1980s, Kuhn kept referring to "the more theoretical analysis on which I am currently engaged" (Kuhn 1981/1987, p. 7) and to a "projected book" (Kuhn 1991a, p. 3). The book was intended to address the philosophical problems which had been opened with *Structure*, most notably concerning realism and truth, but also such issues as rationality and relativism. Although the projected book remained unfinished, a series of papers from the late 1980s and early 1990s (Kuhn 1989a, 1990, 1991a, 1992, 1993) indicate the direction his work was taking.

On a number of issues his views seemed to have changed in important ways since the publication of *Structure*, most notably views regarding the family resemblance nature of concepts (chapter 6.1), the function of incommensurability in the development of science (chapter 6.2), and the importance of history for the philosophy of science (chapter 6.3).

6.1 Concepts

One of the central issues in Kuhn's work since *Structure* was the development of a family resemblance account of concepts (see chapter 4.2). Since the 1970 postscript to *Structure* Kuhn had claimed that all concepts, including scientific concepts, are based on similarity rather than definition. However, in the last paper he published (Kuhn 1993), he introduced a distinction between what he called nomic and normic concepts; a distinction which may be seen as the reaction to problems inherent in a pure family resemblance account of scientific concepts (chapter 6.1.3). However, for both kinds of concepts Kuhn claimed that as kind terms they serve a specific function that can be traced to the evolution of certain neural mechanisms (chapter 6.1.2)

6.1.1 Normic and nomic concepts

In his 1993 "Afterwords", Kuhn introduced a new distinction between *normic* concepts and *nomic* concepts based on whether or not the generalizations in which the concepts form part are exceptionless.[1] Normic concepts allow for exceptions in the generalizations usually satisfied by the referents. For example, the normic concepts 'liquid', 'gas' and 'solid' are involved in generalization such as "liquids expand when heated". This is a generalization which may sometimes fail, for example for water between 0 and 4 degrees centigrade (cf. Kuhn 1993, p. 316). Nomic concepts, on the contrary, are concepts for which the generalizations are exceptionless laws of nature. For example, the concept 'force' is a nomic concept involved in generalizations such as Newton's three laws of motion, which are all exceptionless laws of nature (cf. Kuhn 1993, p. 316ff).

The distinction between normic and nomic terms bears resemblance to a previous distinction in Kuhn's work between concepts "applied by direct inspection" and concepts for which "laws and theories also enter into the establishment of reference" (Kuhn 1979b, p. 412).[2] The concepts which are applied by direct inspection, or, as Kuhn also called them, the 'basic terms', are learned through ostension. These are concepts like 'duck', 'goose' and 'swan' which are acquired together in contrast sets on the basis of similarity and difference between their instances (see chapter 4.2). Similarly, on Kuhn's view, the concepts for which laws and theories also enter into the establishment of reference, or, as Kuhn also called them, the 'theoretical terms', are learned by having problem situations pointed out to which a given law applies. For example, to teach students the concepts 'force' and 'mass', the instructor may point out problem situations to which Newton's second law apply, such as, for example, the simple pendulum, free fall, or harmonic oscillators. Like for the referents of 'basic terms' Kuhn claimed that what such problem situations "have in common is not that they satisfy some explicit or even some fully discoverable set of rules and assumptions that gives the tradition its character and its hold upon the scientific mind. Instead, they may relate by resemblance" (Kuhn 1970a, p. 45). Hence, prior to his 1993 paper, Kuhn had emphasized the similarity between the two kinds of concepts rather than their difference.

In introducing the distinction between normic and nomic concepts, Kuhn now emphasized the difference in the way the two kinds of concepts are learned. Normic concepts are learned in contrast sets, and Kuhn noted that "the ability to pick out referents for any of these terms depend critically upon the characteristics that differentiate its referents from those of the other terms in the set, which is why the terms involved must be learned together and why they collectively constitute a contrast set" (Kuhn 1993, p. 317). These characteristics of normic concepts are the same as those which Kuhn on the basis of his family resemblance account of concepts had claimed to hold for concepts generally (see chapter 4.2). Further, the core of his family resemblance account had been the claim that various instances of a concept need not all satisfy the same set of criteria. Likewise, introducing the normic/nomic distinction Kuhn emphasized that instances of a normic concept need not always satisfy the same generalizations – the generalizations in which normic concepts form part "are normic, admit exceptions" and "when terms are learned together in this way [by ostension and in contrast sets], each comes with attached normic generalizations about the properties likely to be shared by its referents" (Kuhn 1993, p. 317).

Nomic concepts, on the contrary, are not learned in contrast sets. As Kuhn noted, nomic concepts, such as the concept 'force' are not normally in any contrast set at all (cf. Kuhn 1993, p. 317). However, Kuhn still argued that these concepts cannot be learned individually. They need to be learned together with other terms with which they are closely related in another way than by contrast. Kuhn claimed that a concept like 'force' "must be learned with terms like 'mass' and 'weight'. And they are learned from situations in which they occur together, situations exemplifying laws of nature" (Kuhn 1993, p. 317).

This may seem as if Kuhn's normic/nomic distinction was an attempt to introduce a distinction between similarity class (or family resemblance) concepts and non-similarity class concepts, *viz*. concepts explicitly defined by scientific laws. However, in Kuhn's earlier account of concepts for which laws and theories also enter into the establishment of reference, he focused on the family resemblance between the problem situations to which a given law could be applied. Hence, the difference between normic and nomic concepts may be seen not as a distinction between family resemblance concepts and concepts which can be explicitly defined, but rather as a difference between the level on which family resemblance enters and the complexity of the ostensive acts by which concepts are introduced.

Whereas for normic concepts, in order to learn the concept, several instances of each individual concept in the contrast set are ostended, for nomic concepts what is pointed out are not instances of *individual* concepts but complex problem situations to which a given law applies and which involve the simultaneous use of *several* concepts. For example, for nomic concepts instances of contrasting concepts like 'goose', 'swan', and 'duck' are ostended individually, while for nomic concepts what is pointed out are instances of the application of a natural law like, for example, Newton's second law $F=ma$ in which the concepts 'force', 'mass', and 'acceleration' are simultaneously involved. Previously, Kuhn had focused on the family resemblance between the problem situations to which a given law could be applied, emphasizing the similarity between normic and nomic concepts. Although he later came to emphasize the difference between nomic and normic concepts, he never offered a general account of how the referents of the *individual* nomic concepts such as 'force', 'mass', and 'acceleration' involved in such non-contrasting relations as Newton's second law should be identified.[3]

6.1.2 The biological component of the lexicon

Kuhn's theory of concepts was an important part of his argument against realism. On Kuhn's view, it is not language that must be accommodated to the world, but the world is a product of a mutual accommodation between experience and language (see chapter 5.1). A lexicon gives the members of the community that uses it access to a world which is perceptually and conceptually subdivided in a certain way. Kuhn therefore claimed that "like the Kantian categories, the lexicon supplies preconditions of possible experience" (Kuhn 1991a, p. 12). Hence, on Kuhn's view, the lexicon should not be seen as a set of beliefs, but as "a mental module prerequisite to having beliefs, a mode that at once supplies and bounds the set of beliefs it is possible to conceive" (Kuhn 1991a, p. 5).

Kuhn expected this mental module to be both biologically and culturally determined: "Doubtless some aspects of [the] lexical structure are biologically determined, the products of a shared phylogeny. But, at least among advanced creatures (and not just those linguistically endowed), significant aspects are determined also by education" (Kuhn 1991a, p. 10).[4] The biological component made Kuhn reflect on the evolutionary origin of the module, claiming that "presumably it evolved originally for the sensory, most obviously for the visual, system" and that it developed from "a still more fundamental mechanism which enables individual living organisms to reidentify other substances by tracing their spatio-temporal trajectories" (Kuhn 1991a, p. 5).

6.2 The evolution metaphor

Kuhn's evolutionary epistemology had its origin in *Structure* where he argued that the historical development of the sciences is a development onwards *from* a given point in history rather than a teleological development towards a well-defined end point (see chapter 3.2). This led Kuhn to compare the development of science with biological evolution: "scientific development is, like biological, a unidirectional and irreversible process" (Kuhn 1970a, p. 206).

Kuhn originally saw the parallel between the scientific and biological development in the roles played by theory choice and natural selection: "Verification is like natural selection: it picks out the most viable among the actual alternatives in a particular historical situation" (Kuhn 1970a, p. 146, similarly p. 173). On this view, what happens when a scientific community has been brought to a state of crisis and extraordinary research brings the proliferation of alternative theories, diverging more and more from the previously accepted theory (see chapter 3.2.4) can be seen as some sort of multiple 'mutations' of the theory of which only the most promising will finally survive.

At the beginning of the 1990s Kuhn changed his view of the parallel between the development of science and biological evolution, now claiming that "the biological parallel to revolutionary change is not mutation, as I thought for many years, but speciation" (Kuhn 1991a, p. 8). Kuhn introduced the speciation metaphor to describe the historical fact that the development of science has produced more and more scientific specialties. According to Kuhn this proliferation of specialties can happen in two different ways. Either a new specialty may split off from its parent, as happened when, for example, computer science split off from mathematics, or a new specialty may be formed in the area of overlap between existing specialties, as happened when, for example, physical chemistry and biochemistry arose as distinct research specialties. It may seem as if only the former kind of development leads to a proliferation of research specialties whereas the latter is a development towards an increasing unification of science. However, Kuhn emphasized that the emergence of new specialties in the area of overlap between existing specialties does not lead a unification of these specialties. On the contrary, rather than becoming part of the disciplines from which they originated, these new disciplines become more and more separated from both of their parents, developing their own journals, societies, and departments (cf. Kuhn 1991a, p. 7; 1992, p. 15f.). Hence, Kuhn claimed that "over time a diagram of the evolution of scientific fields, specialties, and sub-specialties comes to look strikingly like a layman's diagram for a biological evolutionary tree" (Kuhn 1991a, p. 7f.).

This may not seem very different from the view he expressed in *Structure* when he claimed that "the net result of a sequence of such revolutionary selections, separated by periods of normal research, is the wonderfully adapted set of instruments we call modern scientific knowledge. Successive stages in that developmental process are marked by an increase in articulation and specialization" (Kuhn 1970a, p. 172). However, in some respects the new metaphor for the development of science in terms of speciation is very different from the metaphor of revolutions introduced in *Structure*.

In *Structure* Kuhn had introduced the notion of revolutions to denote episodes in the history of science in which an older theory was replaced by an incompatible new one (see chapter 3.2.5). In introducing the speciation metaphor Kuhn admitted that this destructive element is "not nearly so directly present in biological speciation" (Kuhn 1992, p. 19). The historical cases which Kuhn drew on to illustrate his views were therefore very different for the two metaphors. In *Structure* he illustrated the notion of revolutions by, for example, the Copernican revolution (Kuhn 1970a, p. 115f.) or the chemical revolutions (Kuhn 1970a, p. 118). His speciation metaphor, on the contrary, is illustrated by quite different cases such as, for example, the separation of various components of mathematics during the 17th century (Kuhn 1992, p. 17), or the development of molecular biology (Kuhn 1991a, p. 7). Although the illustrations of the speciation metaphor are not elaborated in any detail in his work, it is clear that examples like these do not imply the same competition to survive between the old and the new specialty as that which is seen in scientific revolutions.

The most striking difference is the role played by incommensurability in scientific revolutions and speciation of scientific specialties, respectively. According to Kuhn, in the speciation process, the mutual isolation of the subspecialties is brought about by a growing *conceptual disparity* between the developed tools, hence, the specialized scientists with their highly adapted tools, refined to serve the purposes of the subspecialty, inhabit a *niche* isolated from the niches of other subspecialties (cf. Kuhn 1991a,p. 10f.; Kuhn 1992, p. 19f). This conceptual disparity Kuhn ascribed to incommensurability:

> ... what makes these specialties distinct, what keeps them apart and leaves the ground between them as apparently empty space ... is incommensurability, a growing conceptual disparity between the tools deployed in the two specialties. Once the two specialties have grown apart, that disparity makes it impossible for the practitioners of one to communicate fully with the practitioners of the other. (Kuhn 1992, p. 19f.)

However, the conceptual disparity between two specialties placed at different branches of the evolutionary tree of the sciences is very different from the conceptual disparity between the two specialties at each side of a revolutionary divide. To elaborate on Kuhn's speciation metaphor, one could say that the fact that there is no communication between the different niches – such as evolutionary biology and molecular biology – reflects only that they address 'something different', that as with astrophysics and biochemistry, so they are not 'about the same thing'. On the contrary, for such theories as, for example, oxygen theory and phlogiston theory one would say that they are indeed 'about the same thing' and therefore within this shared niche compete on offering the better account of their common domain. When formulated in terms of the speciation metaphor this amounts to the difference that scientists inhabiting some given scientific niche live in peaceful co-existence with scientists from *other niches*, but competition and combat is the immediate result if intruders, or betrayers from inside, start exploiting their *own niche*, changing it slightly as they go. Claiming incommensurability to hold both between different (sub)specialties and between a new subspecialty and an old one whose problems the new is expected to solve, Kuhn therefore seemed to blur this distinction between intruders in a given niche and inhabitants of another.

6.3 Historical evidence or first principles

During the 1960s, Kuhn had been one of the key figures in the development of a historical philosophy of science (see chapter 2.3). As the historically inclined philosopher of science Mary Hesse pointed out in her review of *Structure* for the history of science journal *Isis* in 1963 "it cannot be disputed that this is the first attempt for a long time to bring historical insights to bear on the philosophers' account of science" (Hesse 1963, p. 287). Nearly thirty years later, celebrating Kuhn's work at a conference at the MIT, the philosopher Michael Friedman maintained that "Thomas Kuhn's *The Structure of Scientific Revolutions* (1962) forever changed our appreciation of the philosophical importance of the history of science" and later that "it is clear beyond the shadow of a doubt, I think, that careful and sensitive attention to the history of science must remain absolutely central in any serious philosophical consideration of science" (Friedman 1993, p. 37).

Nevertheless, in some of his last writings, Kuhn began modifying the emphasis on the importance of history:

> "When I first got involved, a generation ago, with the enterprise now often called historical philosophy of science, I and most of my coworkers thought history functioned as a source of empirical evidence. That evidence we found in historical case studies, which forced us to pay close attention to science as it really was. Now I think we overemphasized the empirical aspect of our enterprise (an evolutionary epistemology need not be a naturalized one). What has for me emerged as essential is not so much the details of historical cases as the perspective or the ideology that attention to historical cases bring with it." (Kuhn 1991a, p. 6, similarly 1992, p. 4).

The perspective which Kuhn found essential is a *developmental* view of science. On this view, science is not seen as a static body of knowledge, but as an ever developing enterprise (cf. Kuhn 1992, p. 10). Consequently, the interesting questions for the philosopher of science to address are not about the rational warrant for taking a particular belief to be true, but the rational warrant for belief *change*.

As Kuhn saw it, this view would simply dismiss a number of the problems that had haunted the philosophy of science. First, the theory-dependence of observation ceases to pose an obstacle for rationality: "For the historical perspective ... the *rationality* of the conclusions requires only that the observations invoked be neutral for, or shared by, the members of the group making the decision. ... By the same token, the observations involved need no longer be independent of all prior beliefs but only of those that would be modified as a result of the change" (Kuhn 1992, p. 11f.).

Second, evaluating beliefs for their truth usually involves a set of secondary criteria such as accuracy, consistency, simplicity or range of application which have all proven highly equivocal. On the contrary, Kuhn claimed that "to ask which of two bodies of belief is *more* accurate, displays *fewer* inconsistencies, has a *wider* range of applications, or achieves these goals with the *simpler* machinery does not eliminate all ground for disagreement, but the comparative judgement is clearly far more tractable than the traditional one from which it derives" (Kuhn 1992, p. 13).

The price that has to be payed for accepting the developmental view is to give up the realist assumption of the existence of a fixed realm of theory-independent entities (see chapter 5.1) as well as the correspondence theory of truth (see chapter 5.2). However, as Kuhn saw it, in adopting a developmental perspective, the conclusions on theory choice, truth and realism which had previously been drawn from the historical record could "be derived instead from first principles. Approaching them in that way reduces their apparent continency, making them harder to dismiss as a product of muckraking investigation by those hostile to science" (Kuhn 1992, p. 10). Just what the nature of these "first principles" would be remained an unsolved problem in Kuhn's philosophy of science.

Notes

PREFACE

[1] The manuscript of Kuhn's Lowell Lectures can be found in Thomas S. Kuhn Papers, (MC 240), MIT Institute Archives and Special Collections, MIT Libraries, Cambridge, Massachusetts.

1. BIOGRAPHICAL SKETCH

[1] This chapter builds on information provided by Kuhn in interviews (Kuhn et al. 1997 and Sigurdsson 1990) and in his writings as well as on historical work by Andresen (1999) and Hufbauer (forthcoming).

[2] He had also attended a few of Sarton's lectures in the history of science, but had found them "turgid and dull" (Kuhn et al. 1997, p. 161).

[3] This experience was turned into a maxim which he offered his students: "When reading the works of an important thinker, look first for the apparent absurdities in the text and ask yourself how a sensible person could have written them. When you find an answer ... when those passages make sense, then you may find that more central passages, ones you previously thought you understood, have changed their meaning" (Kuhn 1977a, p. xii).

[4] Hence, while his historical interest was not in historical details as such, he later recalled that "I could read texts, get inside the heads of the people who wrote them, better than anybody else in the world. I loved doing that. I took real pride and satisfaction in doing it. So being a historian of *that* sort was something I was quite willing to do and got a lot of kicks out of being, and did my best to teach other people to do" (Kuhn et al. 1997, p. 162).

[5] For an account of Kuhn's influence on the social sciences, see e.g. Barnes 1982 or Gutting 1980.

Notes

⁶ For a treatment of the relation between history and philosophy of science in Kuhn's work, see e.g. Caneva 2000.
⁷ An annotated edition of the manuscript edited by John Conant and John Haugeland is planned to be published by University of Chicago Press.

2. PHILOSOPHICAL CONTEXT

¹ Among the philosophers and scientists who took part in the Circle's activities were Rudolf Carnap (1891-1970), Herbert Feigl (1902-1988), Otto Neurath (1882-1945), Friedrich Waismann (1896-1959), Edgar Zilsel (1891-1944), Victor Kraft (1880-1975), Philipp Frank (1884-1966), Hans Hahn (1879-1934), Kurt Gödel (1906-1978), Karl Menger (1902-1985), and Gustav Bergmann (1906-1987).
² Members of the Berlin Society included, among others, Hans Reichenbach (1891-1953), Carl Gustav Hempel (1905-1997), Richard Martin Edler von Mises (1883-1953), and Kurt Grelling (1886-1942).
³ In the same interview Kuhn reported that the logical positivist literature he had read early in his career was Bridgman's *The Logic of Modern Physics*, "some" Philipp Frank, "some" von Mises and "a little bit of Carnap, but not the Carnap that people later point to as the stuff that has real parallels to me" (Kuhn et al. 1997, p. 186).
⁴ Detailed treatments of the parallel between Kuhn and the late Carnap can be found in Axtell 1993, Earman 1993, Irzik & Grúnberg 1995, and Reisch 1991.
⁵ In his two surveys on the development of history of science and its relations to the philosophy of science, Kuhn mentioned Cassirer only for his historical work (Kuhn 1968a/1977a, p. 108; 1971c/1977a, p. 149).
⁶ For a comparison of Bachelard, Canguilhem and Foucault with Kuhn and other Anglo-Saxon philosophers, see Gutting 1989.
⁷ For the development of the distinction as well as Kuhn's objections against it, see Hoyningen-Huene 1987 and Hoyningen-Huene 1993, ch. 7.4.c.
⁸ For Kuhn's later divergence from the historical approach, see chapter 6.3.

3. THE STRUCTURE OF SCIENTIFIC REVOLUTIONS

[1] A third edition of *Structure* published in 1996 differs from the second edition only by including an index. Unless otherwise indicated, page numbers in this chapter refers to Kuhn 1970a.

[2] On the discussion of historical philosophy of science as descriptive or normative, see chapter 2.3.

[3] For an account of the difference between anachronic (or 'whig') historiography of science and diachronic historiography of science, see e.g. Kragh 1987.

[4] Kuhn's use of these words should not be taken to imply any negative connotations. On the contrary, mop-up work is "fascinating" (p. 24). Consequently, scientists display "enthusiasm and devotion" (p. 36) for the problems of normal research.

[5] It is important to note that in talking about the rules governing puzzle-solving Kuhn had in mind a considerably broadened use of the notion 'rule' which encompasses 'established viewpoints' or 'preconceptions', cf. p. 39. We will return to the notion of rules and Kuhn's rejection of definition-like rules in chapters 3.2.2 and 4.

[6] One notes a slight difference between these two uses of the term: in the first case, an exemplar is an exemplary problem solution while in the second case an exemplar is a classic book in which the accepted examples initially appeared. Kuhn noted this difference in the preface to his later collection of papers, *The Essential Tension* (1977a), p. xix.

[7] When referring to this entire global set of commitments, Kuhn sometimes used the term "paradigm", sometimes the term "theory". Kuhn pointed out that this usage of the term theory is similar to its usage among scientists, but different from its standard usage among philosophers: "Scientists themselves would say they share a theory or set of theories, and I shall be glad if the term can ultimately be recaptured for this use. As currently used in philosophy of science, however, 'theory' connotes a structure far more limited in nature and scope than the one required here" (p. 182). The philosophical view which Kuhn here referred to is most likely the view first advanced by Campbell (1880-1949) and later by the logical

Notes

positivists of scientific theories as a system of axioms and a system of semantic rules for their interpretation.

[8] For a more detailed account of this view, see chapter 4.

[9] On Wittgenstein's and Kuhn's family resemblance accounts of concepts, see chapter 4.2.

[10] A further discussion of the difference between Kuhn and Popper on this point can be found in Kuhn 1970b, Kuhn 1970c and Popper 1970.

[11] This disagreement about foundations suggests a certain similarity between extraordinary research and pre-paradigmatic research. However, there is an important difference between these two kinds of research since extraordinary research is focused on solving specific recalcitrant problems and is therefore much more well-defined and restricted in scope than pre-paradigmatic science (cf. p. 84).

[12] Advancing a phase model for the development of science according to which science develops through successive periods of tradition-preserving normal science and tradition-shattering revolutions, Kuhn's *Structure* has often been compared to Fleck's *Genesis and Development of a Scientific Fact* (Fleck 1935/1979) and to Foucault's *The Order of Things* (Foucault 1966/1970). Drawing on a case study from medicine on the development of a diagnostic test for syphilis, Fleck (1896-1961) argued that a community, a 'thought collective', forms a functional unit in which people interact intellectually, tied together through a certain 'thought style' that forces narrow constraints upon the thinking of the individual. Similarly, drawing on an analysis of the development of the human sciences from the Renaissance to the present, Foucault (1926-1984) advanced the view that during this period four different epistemes can be recognized. Each of these epistemes determined the conditions for all knowledge of their time, and the transition from one episteme to the next happens as a break that imply radical changes in the conception of knowledge (see chapter 2.2). Comparisons between Kuhn's paradigms, Fleck's thought styles, and Foucault's epistemes may be potentially fruitful, but it is easy to be led astray by the apparent similarity and overlook the many differences between their positions. Detailed comparisons of Kuhn and Foucault can be found in Hacking 1979 and Weinert 1982. Comparisons of Kuhn and

Fleck can be found in Brorson & Andersen (forthcoming), Cohen & Schnelle 1986, and Harwood 1986.
[13] For a detailed treatment of this history, see Cohen 1985.
[14] Kuhn's insistence on scientific revolutions as sudden and unstructured events developed considerably over the years. In *Structure* his main emphasis in describing the transition was on perception, drawing heavily on the psychological notion of gestalt swithes. During the late 1970s and early 1980s when his emphasis changed to describing revolutions as a change in language, he still insisted that linguistic drift is prohibited (cf. 1983b, p. 715). However, simultaneously he admitted that the gestalt switch metaphor derived from his experiences as a historian working backwards in time, whereas the experiences of scientists moving through time in the opposite direction might be different. For a further discussion of this development, see chapter 4.4.2.
[15] In its standard mathematical usage, "incommensurability" means "no common measure". For example, the circumference of a circle is incommensurable with its radius in the sense that there is no unit of length that may be contained an integral number of times in each without residue.
[16] Kuhn's first use of the term 'incommensurable' in *Structure* is in the introduction where he describes the competing schools of a pre-paradigmatic science by "what we shall come to call their incommensurable ways of seeing the world and of practicing science in it" (p. 4).
[17] On this particular aspect of incommensurability, see also chapter 4.4.
[18] On this particular aspect of incommensurability, see also chapter 5.1.
[19] For an account of the change of emphasis in Kuhn's work from perception to linguistics, see e.g. Hoyningen-Huene 1993, chapters 3.5 and 3.6.
[20] This is a point which was repeatedly stressed by Kuhn in his subsequent work, see e.g. Kuhn 1970c, p. 250; 1983a, p. 670f.; 1991a, p. 4

4. SCIENTIFIC CONCEPTS

[1] Various aspects of Kuhn's views on science education have been closely analyzed in the journal *Science & Education*

which devoted two full issues to the topic 'Thomas Kuhn and science education', *Science & Education* 9(1-2), 2000 (edited by Michael Matthews).

[2] Further, it is important to note that although Kuhn argues that dogmatism is required for normal science, and that the kind of science education described above does lead to such dogmatism, he does *not* claim that this is the *only* possible way in which science education can take place. Still, his aim was to argue against the then (and now) popular educational theory that emphasizes divergent thinking and the freedom to go off in different directions, suggesting instead that "an educational system best described as an initiation into an unequivocal tradition should be thoroughly compatible with successful scientific work" (Kuhn 1959a/1977a, p. 237).

[3] The procedure by which science students are supposed to model novel problems to the exemplary problems is obviously very close to the procedure by which language students learn to conjugate e.g. Latin verbs by reciting *amo, amas, amat, amamus, amatis, amant*. Hence, Kuhn adopted the expression for such standard examples in language teaching, 'paradigm', and simply extended it to cover standard examples in science teaching as well.

[4] For example, in the dialogue *Euthyphro*, Socrates asks Euthyphro to tell him what the pious is. When Euthyphro initially answers by giving examples of pious actions, Socrates maintains that he did not ask to be told some of the pious actions but the form itself which makes the pious actions pious (Plato 1981).

[5] This view draws on the distinction originally introduced in the 17th century in the treatise on logic published by the French school of Port-Royal between the extension as the class of objects falling under a given concept and the intension as the list of features which all object falling under the concept share (Arnauld & Nicole 1662/1996, section I 6).

[6] This claim Kuhn saw as closely connected to the rejection of the analytic-synthetic distinction that Quine had argued for (Quine 1951/1953). See Kuhn 1970a, p. vi, Kuhn 1961a/1977a, p. 186, or Kuhn 1964/1977a, p. 258.

[7] In his published work Kuhn never referred to the literature on children's concept acquisition, but drew solely on common everyday experience. For the similarity between Kuhn's

account and the account developed by cognitive psychologists on the basis of extensive experiments on concept learning and categorization, see e.g. Andersen, Barker & Chen 1996 and Nersessian 1998.

[8] The only example of the acquisition of scientific concepts which Kuhn spelled out in some detail is his analysis of how students learn the concepts force, mass and weight, see Kuhn 1989a, pp. 15-21 and Kuhn 1990a, pp. 301-308.

[9] Kuhn later changed his view, separating concepts like 'mass' and 'force' from pure family resemblance concepts. On this change, see chapter 6.3.

[10] For an overview of this development, see e.g. Gardner 1987.

[11] The literature on Wittgenstein's notion of family resemblance is vast, but a useful survey of the particular problem of the possible limitation of concepts' extensions can be found in Bellaimey 1990.

[12] A more detailed reconstruction of the argument can be found in Andersen 2000.

[13] Analysis of hierarchical conceptual systems formed by subdivision have been offered in treatments of logic since antiquity (see also chapter 4.4.3). Kuhn's restriction of dissimilarity to concern only contrasting concepts can be found within other fields working on similarity, such as the cognitive psychology (e.g. Rosch 1987, p. 157), or ethnographic semantics and cognitive anthropology (e.g. Conklin 1969, Kay 1971). For a discussion of Kuhn's view of hierarchical conceptual systems compared to views of other scholars, see further chapter 4.4.3.

[14] In his published work Kuhn only briefly mentioned that he saw the basis of taxonomies in a "fundamental mechanism which enables individual living organisms to reidentify other substances by tracing their spatio-temporal trajectories" (Kuhn 1991a, p. 5). On this view, part of a taxonomy will be "biologically determined, the product of a shared phylogeny" (Kuhn 1991a, p. 10). See also chapter 6.1.2).

[15] Interest in conceptual hierarchies can be traced back to Plato's method of division, that is, the method of definition which works by bipartition of concepts from the most general and downwards, and to Aristiotle's treatment of definitions in terms of genus and differences which in the scholastic tradi-

tion was developed into definitions *per genus proximum and differentiam specificam*.

[16] On the difficulties in distinguishing between severe and non-severe anomalies and on the possible outcomes of extraordinary research, see chapters 4.2.3 and 4.2.4.

[17] Nersessian and Andersen (1997) offers an account of how a typology of changes and correlations between specific kinds of changes and certain kinds of anomalies can be derived from the three taxonomic principles.

[18] For an analysis of the conceptual structures of Aristotelian and Newtonian mechanics, see e.g. Nersessian 1989, Nersessian & Resnick 1989.

5. PHILOSOPHICAL IMPLICATIONS

[1] This collection includes criticism of Kuhn's views by, among others, Toulmin, Popper and Feyerabend. Another important collection discussing the ideas advanced in *Structure* is Gutting 1980. Recent collections on the work of Thomas Kuhn include Horwich 1993 and *Configurations* 6, no. 1, 1998 (Special Issue on Thomas Kuhn, edited by N.J. Nersessian). Extensive bibliographies on discussions of Kuhn's work can be found in Gutting 1980 and Hoyningen-Huene 1993.

[2] For an exposition of various realist replies to Kuhn's thesis, see e.g. Sankey 1994. For a collection of recent views by some of the main participants in the debate, see e.g. Sankey & Hoyningen-Huene (forthcoming).

[3] This theory was originally introduced by Kripke to cover proper names (Kripke 1972). Putnam adopted it first to cover physical magnitude terms and extended it to cover natural kind terms in general (Putnam 1973, 1975). Other scholars who have used the causal theory of reference to argue against Kuhn's incommensurability thesis include Devitt (1984), Kitcher (1978), Nola (1980) Sankey (1994) and Sankey (1997).

[4] Kuhn's view of a Kantian world-in-itself seem to have changed over the years. Responding to the causal theorist Boyd's view of how science zero in on nature's real joints (Boyd 1979/1993), Kuhn responded that "Boyd's world with its joints seem to me, like Kant's 'things in themselves', in

principle unknowable. The view toward which I grope would also be Kantian but without 'things in themselves'" (Kuhn 1979b, p. 419b). But a decade later he stated that "... lexical categories, unlike their Kantian forebears, can and do change, both with time and with passage from one community to another. ... Underlying all these processes of differentiation and change, there must, of course, be something permanent, fixed, and stable. But, like Kant's *Ding an sich*, it is ineffable, undescribable, undiscussible. Located outside of space and time, this Kantian source of stability is the whole from which have been fabricated both creatures and their niches, both the 'internal' and the 'external' worlds" (Kuhn 1991a, p. 12). However, it must be noted that whereas in the former case he rejects the idea of a multitude of things-in-themselves, in the latter case he adopts the idea of a single world-in-itself.

[5] Because on Kuhn's view the perceptually and conceptually subdivided world was dependent of conceptual structures and would therefore change with a change of language, Kuhn maintained that his view was Kantian but "with categories of the mind which could change with time as the accommodation of language and experience proceed" (Kuhn 1979b, p. 418f.). A decade later, he elaborated on this comparison, now directly comparing Kant's categories with the lexicon: "Like the Kantian categories, the lexicon supplies preconditions of possible experience. But lexicon categories, unlike their Kantian forebears, can and do change, both with time and with the passage from one community to another" (Kuhn 1991a, p. 12). However, it must be noted that whereas Kuhn's lexicons comprise all kind terms (cf. Kuhn 1991a, p. 4), Kant's categories are the most fundamental concepts of thought that enable us to form judgements. These comprise twelve different categories, such as, for example, unity, negation, or causality (cf Kant 1781/1787/1997, pp. 204ff.). Hence, Kant's categories are of a much more general and abstract kind and cannot easily be compared to Kuhn's lexicon beyond the observation that both serve a role (although the two roles are different) in establishing synthetic a priori knowledge. For the synthetic a priori in Kant's and in Kuhn's philosophy, see e.g. Brown (1975).

Notes

[6] Hoyningen-Huene describes the phenomenal world as constituted both by object-sided moments (the world-in-itself) and by subject-sided moments (the paradigm).

[7] The *ability* to inherit a structured perceptual space is dependent on the human perceptual apparatus; an issue which Kuhn especially addressed in the 1990s, see chapter 6.1.2.

[8] It is important that this restructuring might not just imply that what was previously an instance of one concept becomes an instance of another. On Kuhn's view, the restructuring could even change what is considered to be an object – some objects may cease to exist while others may emerge. Kuhn therefore rejected Hacking's nominalist interpretation of his position according to which individuals do not change with a change of paradigm whereas the grouping of individuals into kinds does (cf. Hacking 1993, p. 277. See also Hacking 1983, pp. 109ff.).

[9] Kuhn's only direct reference to a particular theory of truth which he expected would meet his requirements was to the redundance theory of truth (Kuhn 1991a, p. 8f.). The redundance theory of truth has been given many different names, including the deflationary theory, the no-truth theory, the disquotational theory, and the minimal theory. According to this theory, truth does not have a specific nature of which we can develop a theory of truth. To assert that the statement 'sharks have sharp teeth' is true is equivalent to simply asserting that sharks have sharp teeth, and that is all we can say about the truth of the statement 'sharks have sharp teeth'. Thus, according to the redundance theory, to assert that a statement is true is simply to assert the statement itself, and this is all we can say about the truth of the statement.

[10] It is important to note that this new consensus is reached on the background of the same differing interpretations and weighings of the shared values that were previously responsible for dissolving the previous consensus. Kuhn explicitly rejected the view that the new consensus is obtained simply because the interpretations and weighings of the values are changed: "What converges as the evidence changes over time need only be the value [for the probability of a theory T on the evidence E] that individuals compute from their individual algorithms. Conceivably those algorithms themselves also become more alike with time, but the ultimate unanimity of theory choice provides no evidence whatsoever that they do

so. If subjective factors are required to account for the decisions that initially divide the profession, they may still be present later when the profession agrees" (Kuhn 1977, p. 329).

6. LATER DEVELOPMENTS

[1] This distinction has been widely overlooked in the Kuhn literature. For a detailed treatment of the distinction, see Andersen & Nersessian 2000.

[2] This distinction is related to but also different from the logical positivists' distinction between theoretical and observational terms. Kuhn clearly rejected the distinction between observational and theoretical terms on the grounds that "the distinction between a theoretical and a basic vocabulary will not do in its present form because many theoretical terms can be shown to attach to nature in the same way, whatever it may be, as basic terms." (Kuhn 1974/1977a, p. 302, fn. 11). For the relation between Kuhn's distinction and the logical positivists' distinction, see e.g. Hoyningen-Huene 1993, chapter 3.6.a.

[3] Kuhn's treatment of this problem was confined to an analysis of a singe example, the acquisition of the concepts 'force', 'mass' and 'weight. In this analysis Kuhn argued that prior to the students' exposure to Newtonian terminology, other significant portions of the lexicon must already be in place. According to Kuhn, these concepts include "... a vocabulary adequate to refer to physical objects and to their locations in space and time. ... a mathematical vocabulary rich enough to permit the quantitative description of trajectories and the analysis of velocities and accelerations of bodies moving along them. ... a notion of extensive magnitude, a quantity whose value for the whole of a body is the sum of its values for the body's parts" (Kuhn 1989a, p. 15). But further, some of the concepts to be learned may be available in a still-qualitative form. The concept 'weight' is available in the qualitative form referring the particular sort of force which causes a physical body to press on its support while at rest or to fall when unsupported. Likewise, the concept 'quantity of matter', which can be used to introduce the Newtonian concept 'mass', is available as the quantifyable substrate underlying physical bodies (cf. Kuhn 1989a, p. 17). Hence, for both 'weight' and 'mass', their referents can be picked out by the

same qualitative features on both Newtonian and pre-Newtonian usage. However, the Newtonian use is quantitative. According to Kuhn, the Newtonian quantification is learned by exposure to problem situations to which the concepts are applied. Kuhn described two standard routes: one in which students are presented with Newton's Second Law, $F=ma$, as a description of the way moving bodies actually behave, or one in which students are presented with the gravitational law. In both cases, students learn to quantify the notion of force by means of a spring balance and recourse to Hook's Law, $F=-kx$, and using this Newtonian concept of 'force' they can quantify the notions of mass and weight. See also Andersen & Nersessian 2000 for a frame-based analysis of this and other exaples.

[4] Since Kuhn's taxonomies are based on family resemblance, the claim that taxonomic structures are partly biologically determined comes close to views advanced by Quine as part of his naturalist epistemology. Thus, Quine also asserted that "a standard of similarity is in some sense innate. This point is not against empiricism; it is a commonplace of behavioral psychology. (Quine 1969, p. 11). Further, both Kuhn and Quine argued that at least parts of this biological equipment are not only characteristic of humans but of animals as well (cf. Quine 1969, p. 11; Kuhn 1991a, p. 5).

Bibliography

The Works of Thomas S. Kuhn
- (1951a): Newton's "31st Query" and the Degradation of Gold, *Isis* 42: 296-298
- (1952a): Robert Boyle and Structural Chemistry in the Seventeenth Century, *Isis* 43: 12-36
- (1952b): Reply to M. Boas: Newton and the Theory of Chemical Solution, *Isis* 43: 123-124
- (1952c): Note on The Independence of Dencity and Pore-size in Newton's Theory of Matter, *Isis* 43: 364-365
- (1955a): Carnot's Version of "Carnot's Cycle", *American Journal of Physics* 23: 91-95
- (1955b): La Mer's Version of "Carnot's Cycle", *American Journal of Physics* 23: 387-389
- (1957a): *The Copernican Revolution: Planetary Astronomy in the Development of Western Thought*, Cambridge MA: Harvard University Press
- (1958a): The Caloric Theory of Adiabatic Compression, *Isis* 49: 132-140
- (1958b): Newton's Optical Papers, in I.B. Cohen (ed.): *Isaac Newton's Papers and Letters on Natural Philosophy, and Related Documents*, Cambridge: Cambridge University Press, pp. 27-45
- (1959a): The Essential Tension: Tradition and Innovation in Scientific Research, in C.W. Taylor & F. Barron (eds.): *Scientific Creativity: Its Recognition and Development*, New York: John Wiley, pp. 341-354, reprinted in Kuhn 1977a, pp. 225-239
- (1959b): Energy Conservation as an Example of Simultaneous Discovery, in M. Clagett (ed.): *Critical Problems in History of Science*, Madison: University of Wisconsin Press, pp. 321-356, reprinted in Kuhn 1977a, pp. 66-104
- (1960): Engineering Precedent for the Work of Sidi Carnot, in *Acte du IXe Congrès d'Histoire des Sciences*, Barcelona: Association para la historia de la sciencia espanola, 1980 pp. 530-535
- (1961a): The Function of Measurement in Modern Physical Science, *Isis* 52: 161-193, reprinted in Kuhn 1977a, pp. 178-224
- (1961b): Sadi Carnot and the Cagnard Engine, *Isis* 52: 567-574
- (1962a): *The Structure of Scientific Revolutions*, Chicago: University of Chicago Press
- (1962b): Comment on MacKinnon: Intellect and Motive in Scientific Inventors: Implications for Supply, in *The Rate and Direction of Inventive Activity: Economic and Social Factors*, Princeton: Princeton University Press, pp. 379-384
- (1962c): Comment on Siegel: Scientific Discovery and the Rate of Invention, in *The Rate and Direction of Inventive Activity: Economic and Social Factors*, Princeton: Princeton University Press, pp. 450-457
- (1962d): The Historical Structure of Scientific Discovery, *Science* 136: 760-764, reprinted in Kuhn 1977a, pp. 165-177
- (1963): The Function of Dogma in Scientific Research, in A.C. Crombie (ed.): *Scientific Change. Historical Studies in the Intellectual, Social and Techncal Conditions for Scientific Discovery and Technical Invention, from Antiquity to the Present*, London: Heineman, pp. 347-369

Kuhn, T.S. et al. (1963): Discussion (on The Function of Dogma in Scientific Research), in A.C. Crombie (ed.): *Scientific Change. Historical Studies in*

Bibliography

the Intellectual, Social and Techncal Conditions for Scientific Discovery and Technical Invention, from Antiquity to the Present, London: Heineman, pp. 381-395

Kuhn, T. S. (1964): A Function for Thought Experiments, in *L'aventure de la science. Mélanges Alexandre Koyré*, vol. 2, Paris: Herman, pp. 307-334, reprinted in Kuhn 1977a, pp. 240-265

Kuhn, T.S., J.L. Heilbron, P. Forman & L. Allen (1967): *Sources for the History of Quantum Physics. An Inventory and Report*, Philadelphia: American Philosophical Society

— (1968a): The History of Science, *International Encyclopedia of the Social Sciences*, vol. 14, pp. 74-82, reprinted in Kuhn 1977a, pp. 195-126

Kuhn, T.S. et al. (1969): Contributions to the discussion of New Trends in History, *Daedalus* 98: 871-976

Kuhn, T. S. (1969a): Comment on E.M. Hafner: The New Reality in Art and Science, *Comparative Studies in Society and History* 11: 403-412, reprinted as Comments on the Reation of Science and Art in Kuhn 1977a, pp. 340-351

— (1969b): Comment on Folke Dovring: The Principle of Acceleration: A Nondialectical Theory of Progress, *Comparative Studies in Society and History* 11: 426-430

Kuhn, T. S. & J.L. Heilbron (1969): The Genesis of the Bohr Atom, *Historical Studies in the Physical Sciences* 1: 211-290

Kuhn, T.S. (1970a): *The Structure of Scientific Revolutions*, 2nd ed., Chicago: University of Chicago Press

— (1970b): Logic of Discovery or Psychology of Research, in I. Lakatos & A. Musgrave: *Criticism and the Growth of Knowledge*, Cambridge: Cambridge University Press, pp. 1-24, reprinted in Kuhn 1977a, pp. 266-292

— (1970c): Reflections on my Critics, in I. Lakatos & A. Musgrave: *Criticism and the Growth of Knowledge*, Cambridge: Cambridge University Press, pp. 231-278. To be reprinted in Kuhn (forthcoming)

— (1970d): Comment on Westfall: Uneasily Fitful Reflections on Fits of Easy Transmission, in R. Palter (ed.): *The Annus Mirabilis of Sir Isaac Newton 1666-1966*, Cambridge MA: MIT Press, pp. 105-108

— (1970e): Alexandre Koyré and the History of Science, *Encounter* 34: 67-69

— (1971a): Notes on Lakatos, R.C. Buck & R.S. Cohen (eds.): *PSA 1970. In Memory of Rudolph Carnap. Boston Studies in the Philosophy of Science* VIII: 137-146

— (1971b): Les notions de causalité dans le developpement de la physique, in M. Bunge, F. Halbwachs, T.S. Kuhn, J. Piaget & L. Rosenfeld: *Les Théories de la Causalité*, Paris: Presses Universitaeires de France, pp. 7-18. Reprinted in English translation as Concepts of Cause in the Development of Physics in Kuhn 1977a, pp. 21-30

— (1971c): The Relations between History and the History of Science, *Daedalus* 100: 271-304, reprinted in Kuhn 1977a, pp. 127-161

— (1972a): Scientific Growth: Reflections on Ben-David's "Scientific Role". Review of J. Ben-David: *The Scientist's Role in Society: A Comparative Study* (Englewood Cliffs: Prentice-Hall 1971), *Minerva* 10: 166-178

— (1974a): Second Thoughts on Paradigms, in F. Suppe (ed.): *The Structure of Scientific Theories*, Urbana: University of Illinois Press, pp. 459-482, reprinted in Kuhn 1977a, pp. 293-319

Kuhn, T. S. and et al. (1974): Discussion of Second Thoughts on Paradigms, in F. Suppe (ed.): *The Structure of Scientific Theories*, Urbana: University of Illinois Press, pp. 500-517

Kuhn, T. S. (1975): The Quantum Theory of Specific Heats: A Problem in

Bibliography

Professional Recognition, in *Proceedings of the XIV International Congress for the History of Science 1974*, Tokyo: Science Council of Japan, vol. 1, pp. 170-182 and vol. 4, p. 207
— (1976a): Mathematical versus Experimental Traditions in the Development of Physical Science, *The Journal of Interdisciplinary History* 7: 1-31, reprinted in Kuhn 1977a, pp. 31-65
— (1976b): Theory-Change As Structure-Change: Comments on the Sneed Formalism, *Erkenntnis* 10: 179-199. To be reprinted in Kuhn (forthcoming)
— (1977a): *The Essential Tension: Selected Studies in Scientific Tradition and Change*, Chicago: University of Chicago Press (original in German: *Die Entstehung des Neuen. Studien zur Struktur derWissenschaftsgeschichte*, Frankfurt: Suhrkamp 1977).
— (1977b): The Relations between the History and the Philosohpy of Science, in Kuhn 1977a, pp. 3-20
— (1977c): Objectivity, Value Judgment, and Theory Choice, in Kuhn 1977a, pp. 320-339
— (1978): *Black-Body Theory and the Quantum Discontinuity 1894-1912*, Oxford: Clarendon
— (1979a): History of Science, in P.D. Asquith & H.E. Kyburg: *Current Research in Philosophy of Science*, Ann Arbor: Edwards, pp. 121-128
— (1979b): Metaphor in Science, in A. Ortony (ed.): *Metaphor in Science*, Cambridge: Cambridge University Press, pp. 410-419, 2nd ed. 1993, pp. 533-542). To be reprinted in Kuhn (forthcoming)
— (1979c): Foreword to Fleck: *Genesis and Development of a Scientific Fact*, ed. T.J. Trenn & R. Merton, Chicago: University of Chicago Press, pp. vii-xi
— (1980a): The Halt and the Blind: Philosophy and History of Science. Review of C. Howson (ed.): *Method and Appraisal in the Physical Sciences: The Critical Background to Modern Science, 1800-1905* (Cambridge: Cambridge University Press 1976), *British Journal for the Philosophy of Science* 31: 181-192
— (1980b): Einstein's Critique of Planck, in H. Woolf (ed.): *Some Strangeness in the Proportion: A Centennial Symposium to Celebrate the Achievements of Albert Einstein*, Reading: Addison-Wesley, pp. 186-191
Kuhn, T.S. et al. (1980): Open Discussion Following Papers by M.J. Klein and T.S. Kuhn, in H. Woolf (ed.): *Some Strangeness in the Proportion: A Centennial Symposium to Celebrate the Achievements of Albert Einstein*, Reading: Addison-Wesley, pp. 186-191
Kuhn, T. S. (1981): What are Scientific Revolutions?, in L. Krüger, L.J. Daston & M. Heidelberger (eds.): *The Probabilistic Revolution*, vol. 1, *Ideas in History*, Cambridge MA: MIT Press, pp. 7-22. To be reprinted in Kuhn (forthcoming)
— (1983a): Commensurability, Comparability, Communicability, in P.D. Asquith & T. Nickles (eds.): *PSA 1982. Proceedings of the 1982 Biennial Meeting of the Philosophy of Science Association*, vol. 2, pp. 669-688. To be reprinted in Kuhn (forthcoming)
— (1983b): Response to Commentaries [on Commensurablity, Comparability, Communicability], in P.D. Asquith & T. Nickles (eds.): *PSA 1982. Proceedings of the 1982 Biennial Meeting of the Philosophy of Science Association*, vol. 2, pp. 712-716
— (1983c): Reflections on Receiving the John Desmond Bernal Award, *4S Review: Journal of the Society for Social Studies of Science*, 1: 26-30
— (1983d): Rationality and Theory Choice, *Journal of Philosophy* 80: 563-570. To be reprinted in Kuhn (forthcoming)
— (1983e): Foreword to B.R. Wheaton: *The tiger and the shark. Empirical roots of wave-particle dualism*, Cambridge: Cambridge University Press, pp. ix-xiii
— (1984a): Revisiting Planck, *Historical Studies in the Physical Sciences* 14: 231-252
— (1984b): Professionalization Recollected in Tranquility, *Isis* 75: 29-32
Kuhn, T.S. et al. (1985): Panel Discussion on Specialization and Professionalism

Bibliography

within the University, *The American Council of Learned Societies Newsletter* 36 (3/4): 23-27

Kuhn, T. S. (1986): The Histories of Science: Diverse Worlds for Diverse Audiences, *Academe. Bulletin of the American Association of University Professors* 72(4): 29-33

— (1989a): Possible Worlds in History of Science, in S. Allén (ed.): *Possible Worlds in Humanities, Arts, and Sciences*, Berlin: de Gruyter, pp. 9-32. To be reprinted in Kuhn (forthcoming)

— (1989b): Speaker's Reply [to commentaries to Possible Worlds in History of Science], in S. Allén (ed.): *Possible Worlds in Humanities, Arts, and Sciences*, Berlin: de Gruyter, pp. 49-51

— (1990): Dubbing and Redubbing: The Vulnerability of Rigid Designation, in C.W. Savage (ed.): *Scientific Theories, Minnesota Studies in the Philosophy of Science* XIV, Minneapolis: University of Minnesota Press, pp. 298-318

— (1991a): The Road since Structure, PSA 1990(2): 3-13. To be reprined in Kuhn (forthcoming)

— (1991b): The Natural and the Human Sciences, in. D.R. Hiley, J.F. Bohmen & R. Shusterman (eds.): *The Interpretive Turn. Philosophy, Science, Culture*, Ithaca: Cornell University Press, pp. 17-24. To be reprinted in Kuhn (forthcoming)

— (1992): *The Trouble with the Historical Philosophy of Science*, Robert and Maurine Rotschild Distinguished Lecture, 19. november 1991, Cambridge MA: Department of the History of Science, Harvard University. To be reprinted in Kuhn (forthcoming)

– (1993): Afterwords, in P. Horwich (ed.): *World Changes. Thomas Kuhn and the Nature of Science*, Cambridge MA: MIT Press, pp. 311-342. To be reprinted in Kuhn (forthcoming)

Kuhn, T.S., A. Balta, K. Gavrogglu & V. Kindi (1997): A Discussion with Thomas S. Kuhn. A Physicist who became a Historian for Philosophical Purposes, *Neusis* 6: 145-200. To be reprinted in Kuhn (forthcoming)

Kuhn, T.S. (forthcoming): *The Road since Structure*, Chicago: University of Chicago Press

Secondary litterature

Andersen, H. (2000): Kuhn's Account of Family Resemblance: A Solution to the Problem of Wide-Open Texture, *Erkenntnis* 52: 313-337

Andersen, H., P. Barker & X. Chen (1996): Kuhn's Mature Philosophy of Science and Cognitive Psychology, *Philosophical Psychology* 9: 347-363

Andersen, H. & N.J. Nersessian (2000): Nomic Concepts, Frames, and Conceptual Change, *Philosophy of Science* 67(Proceedings): S224-S241

Andresen, J. (1999): Crisis and Kuhn, *Isis* 90(supplement): S43-S67

Arnaud, A. & P. Nicole (1662/1996): *Logic or the Art of Thinking*, ed. J.V. Buroker, Cambridge: Cambridge University Press 1996.

Axtell, G.S. (1993): In the Tracks of the Historicist Movement: Re-assessing the Carnap-Kuhn Connection, *Studies in the History and Philosophy of Science* 24: 119-146

Barnes, B. (1982): *T.S. Kuhn and Social Science*, London: Macmillan

Bellaimey, J.E. (1990): Family Resemblance and the Problem of Under-Determination of Extension, *Philosophical Investigations* 13: 31-43

Boyd, R. (1979/1993): Metaphor and theory change: What is "metaphor" a metaphor for?, in A. Ortony (ed.): *Metaphor and Thought*, Cambridge: Cambridge University Press, pp. 356-408. Revised 2nd edition 1993, pp. 481-

Bibliography

Bridgman, P.W. (1927): *The Logic of Modern Physics*, New York: Macmillan

Brorson, S. & H. Andersen (forthcoming): Stabilizing and Changing Phenomenal Worlds: Ludwik Fleck and Thomas Kuhn on Scientific Literature, to appear in *Journal for the General Philosophy of Science*

Brown, H. (1975): Paradigmatic Propositions, *American Philosophical Quarterly* 12: 85-90

Caneva, H. (2000): Possible Kuhns in the History of Science: Anomalies of Incommensurable Paradigms, *Studies in the History and Philosophy of Science* 31A: 87-124

Churchland, P. (1989): *A Neurocomputational Perspective: The Nature of Mind and the Structure of Science*, Cambridge MA: MIT Press

— (1990): On the Nature of Theories: A Neurocomputational Perspective, in: S.W. Savage (ed.): *Scientific Theories*, Minnesota Studies in the Philosophy of Science, XIV, Minneapolis: University of Minnesota Press, pp. 59-101

— (1992): A Deeper Unity: Some Feyerabendian Themes in Neurocomputational Form, in: R. Giere (ed.): *Cognitive Models of Science*, Minnesota Studies in the Philosophy of Science, XV, Minneapolis: University of Minnesota Press, pp. 341-363

Cohen, I.B. (1985): *Revolution in Science*, Cambridge, MA: Harvard University Press

Cohen, R.S. & T. Schnelle (eds.) (1986): *Cognition and Fact — Materials on Ludwik Fleck*, Dordrecht: Reidel

Conant, J.B. (1947): *On Understanding Science: An Historical Approach*, New Haven, Yale University Press

Conclin, H.C. (1969): Lexicographical Treatment of Folk Taxonomies, in S.A. Tyler (ed.): *Cognitive Anthropology*, New York: Holt, Rinehart & Winston, pp. 41-59

Devitt, M. (1984): *Realism and Truth*, Princeton, N.J.: Princeton University Press

Earman, J. (1993): Carnap, Kuhn, and the Philosophy of Science Methodology, in Horwich (1993), pp. 9-36

Feyerabend, P.K. (1962/1981): Explanation, reduction and empiricism, *Minnesota Studies in the Philosophy of Science*, vol. 3, pp. 28-97. Reprinted in P.K. Feyerabend: *Realism, Rationalism and Scientific Method. Philosophical Papers Volume 1*, Cambridge: Cambridge University Press 1981, pp. 44-96

Fleck, L. (1935/1979): *The Genesis and Development of a Scientific Fact* (ed. T.J. Trenn & R. Merton), Chicago: University of Chicago Press 1979

Foucault, M. (1966/1970): *The Order of Things. An Archaeology of the Human Sciences*, New York: Random House 1970

Friedman, M. (1993): Remarks on the History of Science and the History of Philosophy, in Horwich (1993), pp. 37-54

Gardner, H. (1987): *The Mind's New Science*, New York: Basic Books

Gutting, G. (1980): *Paradigms and Revolutions. Applications and Appraisals of Thomas Kuhn's Philosophy of Science*, Notre Dame: University of Notre Dame Press

— (1989): *Michel Foucault's archaeology of scientific reason*, Cambridge: Cambridge University Press

Hacking, I. (1979): Michel Foucault's Immature Science, *Nous* 13: 39-51

— (1983): *Representing and Intervening. Introductory topics in the philosophy of natural science*, Cambridge: Cambridge University Press

— (1993): Working in a New World: The Taxonomic Solution, in Horwich (1993), pp. 275-310

Hahn, H., O. Neurath & R. Carnap (1929/1973): *The Scientific Conception of the World: The Vienna Circle* in O. Neurath: *Empriricism and Sociology*, edited by M. Neurath & R.S. Cohen, Dordrecht: Reidel 1973, pp. 299-318 (orig. *Wissenschaftliche Weltauffassung: Der Wiener Kreis*, Wien: Arthur Wolf 1929)

Hanson, N.R. (1958): *Patterns of Discovery. An Inquiry into the Conceptual*

Bibliography

Foundations of Science, Cambridge: Cambridge University Press
Harwood, J. (1986): Ludwik Fleck and the Sociology of Knowledge, *Social Studies of Science* 16: 173-187
Hesse, M. (1963): Review of Thomas S. Kuhn: *The Structure of Scientific Revolutions*, *Isis* 54: 286-287
Horwich, P. (ed.) (1993): *World Changes. Thomas Kuhn and the Nature of Science*, Cambridge MA: MIT Press
Hoyningen-Huene, P. (1987): Context of Discovery and Context of Justification, *Studies in the History and Philosophy of Science* 18: 501-515
— (1993): *Reconstructing Scientific Revolutions. Thomas S. Kuhn's Philosophy of Science*, Chicago: University of Chicago Press
Hoyningen-Huene, P. & H. Sankey (eds.) (2001): *Incommensurability and Related Matters*, Dordrecht: Kluwer
Hufbauer, K. (forthcoming): Kuhn's Discovery of History (1940-1958), to appear in *Studies in the Physical and Biological Sciences*
Irzik, G. & T. Grünberg (1995): Carnap and Kuhn: Arch Enemies or Close Allies?, *British Journal for the Philosophy of Science* 46: 285-307
Kant, I. (1781/1787/1997): *Critique of Pure Reason* (trans. & ed. P. Guyer & A.W. Wood), Cambridge: Cambridge University Press 1997
— (1783/1950): *Prolegomena to Any Future Metaphysics* (trans. L.W. Beck), Indianapolis: Bobbs-Merril 1950
— (1800/1992): *Lectures on Logic* (trans. & ed. by J. Michael Young), Cambridge: Cambridge University Press, 1992
Kay, P. (1971): Taxonomy and semantic contrast, *Language* 47: 866-887
Kitcher, P. (1978): Theories, Theorists and Theoretical Change, *Philosophical Review* LXXXVII: 519-547
Koyré, A. (1939/1978): *Galileo Studies*, Hassocks: Harvester Press 1978
Kragh, H. (1987): *An Introduction to the Historiography of Science*, Cambridge: Cambridge University Press
Kripke, S. (1972): Naming and necessity, in G. Harman & D. Davidson (eds.): *The Semantics of Natural Language*, Dordrecht: Reidel, pp. 254-355
Lakatos, I. (1970): Falsification and the Methodology of Scientific Research Programmes, in: Lakatos & Musgrave (1970), pp. 91-196
Lakatos, I. & A. Musgrave (eds.) (1970): *Criticism and the Growth of Knowledge*, Cambridge: Cambridge University Press
Masterman, M. (1970): The Nature of a Paradigm, in: Lakatos Musgrave (1970), pp. 59-89
Minsky, M. & S. Pappert (1969): *Perceptrons. An Introduction to Computational Geometry*, Cambridge MA: MIT Press
Nagel, E. (1961): *The Structure of Science. Problems in the Logic of Scientific Explanation*, New York: Harcourt, Brace & World
Nersessian, N. (1989): Conceptual Change in Science and in Science Teaching, *Synthese* 80: 163-183
— (1998): Kuhn and the Cognitive Revolution, *Configurations* 6: 87-120
Nersessian, N. & H. Andersen (1997): Conceptual Change and Incommensurability: A Cognitive-Historical View, *Danish Yearbook of Philosophy* 32: 111-152
Nersessian, N. & L.B. Resnick (1989): Comparing Historical and Intuitive Explanations of Motion: Does "Naive Physics" Have a Structure?, *Proceedings of the Cognitive Science Society* 11: 412-420
Nola, R. (1980): 'Paradigms Lost, or the World Regained' — An Excursion into Realism and Idealism in Science, *Synthese* 45: 317-350
Plato (1981): *Euthyphro*. In G. Grube (ed. and trans.) *Five Dialogues*, Indianapolis: Hackett Publishing Co. Reprinted in parts in E. Margolis & S. Laurence

Bibliography

(eds.): *Concepts. Core Readings*, Cambridge MA: MIT Press 1999, pp. 87-100

Popper, K.R. (1935/1959): *Logic of Scientific Discovery*, New York: Basic Books 1959

— (1970): Normal Science and its Dangers, in: Lakatos & Musgrave (1970), pp. 51-58

Putnam, H. (1973): Explanation and Reference, in G. Pearce & P. Maynard (eds.): *Conceptual Change*, Dordrecht: Reidel, pp. 199-211. Reprinted in H. Putnam: *Mind, Language and Reality. Philosophical Papers Volume 2*, Cambridge: Cambridge University Press 1975, pp. 196-214

Putnam, H. (1975): The meaning of 'meaning', in K. Gunderson (ed.): *Language, Mind and Knowledge*, Minnesota Studies in the Philosophy of Science, vol. VII, Minneapolis: University of Minnesota Press. Reprinted in H. Putnam: *Mind, Language and Reality. Philosop;hical Papers Volume 2*, Cambridge: Cambridge University Press 1975, pp. 215-271

Quine, W.V.O. (1951/1953): Two Dogmas of Empiricism, *Philosophical Review* 60: 20-43. Revised version in Quine 1953, pp. 20-46

— (1953): *From a logical point of view*, Cambridge MA: Harvard University Press

— (1960): *Word and Object*, Cambridge MA: MIT Press

— (1969): Natural Kinds, in N. Rescher (ed.): *Essays in Honor of Carl G. Hempel*, Dordrecht: Reidel, pp. 5-23

Reisch, G.A. (1991): Did Kuhn Kill Logical Empiricism?, *Philosophy of Science* 58: 264-277

Rosch, E. (1987): Wittgenstein and Categorization Research in Cogntive Psychology, in M. Chapman & R.A. Dixon (eds.): *Meaning and the Growth of Understanding: Wittgenstein's Significance for Developmental Psychology*, Berlin: Springer, pp. 151-166

Rosenblatt, F. (1962): *Principles of Neurodynamics*, Washington DC: Spartan

Sankey, H. (1994): *The Incommensurability Thesis*, Aldershot: Avesbury

— (1997): Incommensurability: The current state of the play, *Theoria* 12: 425-445

Scheffler, I. (1967): *Science and Subjectivity*, Indianapolis: Bobbs-Merrill

Shapere, D. (1964): The Structure of Scientific Revolutions, *Philosophical Review* LXXIII: 383-394

— (1966): Meaning and Scientific Change, in R. Colodny (ed.): *Mind and Cosmos*, Pittsburgh: University of Pittsburgh Press, pp. 41-85. Reprinted in D. Shapere: *Reason and the search for knowledge: investigations in the philosophy of science*, Boston studies in the philosophy of science, vol. 78, Dordrecht: Reidel 1984

Sigurdsson, S. (1990): The Nature of Scientific Knowledge. An Interview with Thomas Kuhn, *Harvard Science Review*, Winter 1990, pp. 18-25

Siegel, H. (1978): Kuhn and Schwab on Science Texts and the Goals of Science Education, *Educational Theory* 28: 302-309

Weinert, F. (1982): Die Arbeit der Geschichte: Ein Vergleich der Analysemodelle von Kuhn und Foucault, *Journal for the General Philosophy of Science* 13: 336-358

Wittgenstein, L. (1953): *Philosophical Investigations*, New York: Macmillan